CURRENT AFRICAN ISSUES 47

I0025481

Natural Resource Governance and EITI Implementation in Nigeria

Musa Abutudu and Dauda Garuba

NORDISKA AFRIKAINSTITUTET, UPPSALA 2011

INDEXING TERMS:
Nigeria
Natural resources
Petroleum industry
Governance
Administrative reform
Institutional framework
International organizations
Corruption
Economic implications

Language checking: Peter Colenbrander
ISSN 0280-2171
ISBN 978-91-7106-708-1
© The authors and Nordiska Afrikainstitutet 2011
Production: Byrå4
Print on demand. Lightning Source UK Ltd.

Contents

LIST OF FIGURES

AGF	Accountant General of the Federation
CBN	Central Bank of Nigeria
CISLAC	Civil Society Legislative Advocacy Centre
COMD	Commercial and Marketing Department (of Nigeria National Petroleum Corporation)
DPR	Directorate for Petroleum Resources
EFCC	Economic and Financial Crimes Commission
ERA	Environmental Rights Action
EITI	Extractive Industries Transparency Initiative
FGD	Focus Group Discussion
FIRS	Federal Inland Revenue Service
G8	Group of Eight (developed countries)
GMD	Group Managing Director
IJVC	Integrated Joint Venture Company
IMTT	Inter-Ministerial Task Team
IOCs	International Oil Companies
JVC	Joint Venture Company
JOA	Joint Operating Agreement
NEITI	Nigerian Extractive Industries Transparency Initiative
NGO	Non-Governmental Organisation
NNPC	Nigerian National Petroleum Company
NSWG	National Stakeholders Working Group
OPTS	Oil Producers Trade Session (at the Lagos Chamber of Commerce and Industry)
PIB	Petroleum Industry Bill
PTDF	Petroleum Trust Development Fund
PWYP	Publish What You Pay
SGF	Secretary to the Government of the Federation
UBA	United Bank for Africa
ZCC	Zero Corruption Coalition

This Current African Issue analyses EITI implementation and the increasing pressure for results in the face of the validation of Nigeria's efforts at transparency and accountability within the context of natural resource governance in Africa. It focuses on three interdependent dimensions of possible interactions and synergies of the EITI mechanism in Nigeria: accountability for revenues from oil and gas as directly linked to government institutions/public policies/mechanisms/aid programmes; the demand side of accountability in terms of the role of civil society; and transparency and anti-corruption processes as they involve the actual practices of public and private companies, including negotiating and implementing ministries, departments and agencies. The study provides a comprehensive analysis of the Nigerian Extractive Industries Transparency Initiative (NEITI), one of the earliest EITI initiatives to be instituted in Africa, and focuses on the oil and gas industry in Africa's largest hydrocarbon producer. NEITI is placed in the context of Nigeria's anti-corruption efforts, particularly as they relate to oil export revenues and government expenditure, and a global initiative aimed at promoting transparency and accountability in natural resource-based economies in the developing world. The authors pay considerable attention to the institutionalisation of NEITI in Nigeria, providing ample up-to-date empirical information and analysis of efforts at promoting transparency and accountability in the oil and gas industry in the country. Focus is placed on the first ever auditing of the Nigerian oil industry in the attempt to institutionalise the process of verifying and matching the volume of exports with payments made, including examining the propriety of the operations of the international oil companies (IOCs) and Nigeria's public oversight institutions. It provides explanations for the wide discrepancies discovered between figures declared by the IOCs and the Central Bank, the poor coordination between government agencies responsible for monitoring and collecting oil revenues, abuses of due process, contradictions within NEITI itself, and enforcement gaps in the NEITI process. Of note in all these is the observation about how political interference and gaps in the statute setting up NEITI limit its scope and effectiveness. The authors also explore the relations and interactions between NEITI and its various stakeholders: civil society, IOCs and government agencies, particularly those charged with regulating the extractive sector (oil). The relationships examined include: NEITI and the oil and gas sector, NEITI and the media, NEITI and the intergovernmental agencies, NEITI and the National Stakeholders Working Group. Others include those between state and IOCs and civil society and the oil workers. This comprehensive analysis of the role and performance of NEITI is followed by an analysis of emerging policy issues. This discussion largely hinges on empowering the NEITI process to be more autonomous and effective in its monitoring, reporting and enforcement roles. The authors take the view that the Petroleum Industry Bill (PIB) under consideration in the National Assembly (which has been an object of contestation between civil society and IOCs seeking to influence its content in their favour), provides a framework for empowering NEITI. In this regard, they critically examine the capacity of civil society to mobilise the public to push for transparency and accountability clauses in the PIB. The study ends with policy recommendations that target the various stakeholders and emphasise the need to make the NEITI board more representative, build the capacity of civil society and the media to popularise the NEITI process and ensure that it serves the interests of the majority of Nigerians, as opposed to those of the ruling elite and IOCs. The study provides rich information on and analysis of the governance of Nigeria's oil sector that will be of great value to policy-makers, scholars and all those interested in the development of Africa's resource-rich countries.

Cyril Obi
Senior Researcher
The Nordic Africa Institute

It has been a while since the Extractive Industries Transparency Initiative (EITI) emerged as a global initiative to promote transparency in payments for oil, gas and minerals by multinationals and the acknowledgement of receipts of such payments by governments in developing countries. EITI aims at promoting a culture of transparency and accountability and at helping poverty reduction and human development in resource-rich countries. However, several analyses of the EITI implementation process suggest that the global initiative lacks the capacity to generate the profound changes required in the complex linkages in the governance of mineral and hydrocarbon exploitation, thus requiring EITI to interact further with relevant national institutions, critical constituencies and mechanisms.

The Nigeria Extractive Industries Transparency Initiative (NEITI) got much local and international support, given how it evolved as part of a comprehensive economic policy reform agenda of former President Olusegun Obasanjo (1999-2007). That support became particularly overwhelming when the NEITI audit report on oil and gas sector covering 1999 to 2004, containing staggering revelations, was released to the Nigerian public in 2006. The report, the first ever major audit of the Nigeria oil and gas industry in the country's history, precipitated great public interest in NEITI and its work. Even though the audit was delayed beyond public expectations, its success was advanced with yet another audit covering 2005, released in September 2009, even as the country awaits the 2006–2008 report. As much as the Nigerian public appreciates the findings and revelations of these reports, concerns have been expressed over the slow implementation of the remedial plans adopted by the Inter-Ministerial Task Team inaugurated for that purpose. Nigeria, having clearly been the flagship of the EITI process by virtue of what it was able to achieve within a short time of signing on to the global EITI, is now a country on a hard road to validation.

It is important to note that in spite of the high expectations within and outside the continent, EITI itself does not have the capacity to generate the profound changes required in the complex chain of governance of mineral and hydrocarbon exploitation in Africa. However, the great potential of EITI is undeniable in ensuring that an inclusive dialogue takes place between the parties most affected by the exploitation of resources, and that this dialogue translates into the accountability of rulers to the people on whose behalf they govern.

ACKNOWLEDGEMENTS

The authors are profoundly grateful to Auwal Musa Rafsanjani and the Civil Society Legislative Advocacy Centre that he directs for providing the opportunity for this research. We also acknowledge Marc Ninerola of Oxfam Intermo, Madeline R. Young and Ekanem Bassey, with whom we worked closely during interview and focus group discussion (FGD) sessions. To all interviewees and participants in our FGDs, who are too many to single out by name, we thank you for your frankness. We deeply acknowledge the comments of Joseph Amenaghawon on the draft and those by the anonymous peer reviewers appointed by the Nordic Africa Institute.

Musa Abutudu and Dauda Garuba

1. INTRODUCTION

NEITI: Contextualising a unique anti-corruption programme

Nigeria was one of the first countries to sign on to the global EITI process, following an official launch in Abuja in February 2004. The launch itself was in fulfilment of an interest earlier indicated by then President Olusegun Obasanjo in November 2003, a year after the initiative was first mooted by former British Prime Minister Tony Blair at the World Sustainable Development Summit held in Johannesburg, South Africa in October 2002. EITI was precipitated by growing concerns about the irreconcilable gap between the quantum of highly prized natural resources exploited in many developing countries and the widespread poverty and underdevelopment in these countries, the majority of whose governments had continued to maintain a veil of secrecy that enabled institutionalised corruption and mismanagement (Garuba 2010). Thus, coming against the backdrop of Nigeria's chequered history marred by mis-governance and outright plunder of its peoples' common wealth, of which oil forms a prominent part, the country's decision to voluntarily accede to the global EITI was widely acclaimed both locally and internationally.

The Nigerian EITI process was premised on the holistic anti-corruption agenda of the Obasanjo administration. The discourse on the "resource curse", which attracted a wave of interest in development economics and political economy circles, may well explain why NEITI, as the Nigerian sub-set of the global EITI is known, was likened to the revenue side of the Obasanjo administration's due process mechanism (Ezekwesili 2006). Nigeria's sense of urgency to sign on to EITI was largely influenced by the findings of a World Bank study commissioned by President Obasanjo's administration in 2000 that revealed "disturbing declines in crude oil output and sales, discrepancies in fund inflows and outflows, weak institutional capacities, and ineffective management of extractive industry revenues" (Garuba and Ikubaje 2010:142). The findings of this study served as the basis for incorporating oil and gas sector reform into the various components of the Obasanjo economic reform programmes. These macroeconomic reforms targeted stabilising the Nigerian economy through improved budgetary planning and implementation, sustained economic diversification and non-oil growth and improved implementation of fiscal and monetary policies. They also aimed at structural reforms focusing on privatisation; civil service, banking sector and trade policy reforms; and governance and institutional reforms anchored on anti-corruption, with all its ramifications for public procurement, public expenditure management and EITI domestication.[1]

---

1. For detailed discussion of each of the reform programmes, see Ngozi Okonjo-Iweala and Philip Osafo-Kwaako, *Nigeria's Economic Reforms: Process and Challenges*, Working Paper No. 6, Brookings Global Economic Development, Brookings Institution, Washington DC, 2007.

Nigeria under former President Obasanjo demonstrated a strong commitment to the EITI implementation process. The working relationship between President Obasanjo and the chair of NSWG of NEITI was one factor that was apparently crucial in generating the internal momentum for the EITI implementation process in Nigeria (Publish What You Pay and Revenue Watch Institute 2006:10). The concerns expressed that Obasanjo's successor might not demonstrate the same commitment to the EITI administration prompted the legislation on NEITI. In passing the law establishing NEITI in 2007, Nigeria became the first country to formally implement the EITI process with an enabling legal framework. Prior to the legislation, the Obasanjo administration took certain practical steps to implement the EITI process in Nigeria by creating a NEITI secretariat headed by Obiageli Ezekwesili, well regarded within the presidency and the donor community for her excellent bureaucratic skills and commitment to due process in budgetary matters. The secretariat launched the first audit of the oil and gas industry in Nigeria in about 50 years. This audit covered the period 1999–2004. A second audit for 2005 has since been conducted, with findings disseminated to the Nigerian public and the international community, while further plans are currently under way for the 2006–08 audit, the Joint Development Zone (JDZ) with São Tomé and Principe,[2] and the first ever EITI audit in the solid minerals sector.

After signing up for EITI, Nigeria established a NEITI secretariat and constituted the multi-stakeholder NSWG with responsibility for overall policy formulation and supervision of the EITI process in Nigeria. It was these institutions and processes that were formalised in a legal framework provided under the 2007 NEITI Act. By so doing, the revenue end of the fiscal chain for the first time got its own specific anti-corruption attention. It was on this legal basis that the 1999–2004 and 2005 audits were accorded recognition and profoundly celebrated as milestones. How far Nigeria has sustained these achievements is a subject for further review in another section of this study.

---

2. The Joint Development Zones became a major trend in international law governed under Articles 74(3) and 83(3) of the United Nations Convention on the Law of the Sea adopted on 12 December 1982.The provisions allow nation states to contemplate and adopt "provisional arrangements" of "a practical nature" in the event of deadlocks in negotiations over disputed maritime delimitation for a transitional period, while remaining under the duty of carrying on negotiations. The Nigeria-São Tomé and Principe JDZ was adopted in light of the provisions of the law when the parties reached a deadlock over the delimitation of their exclusive economic zone in 2000. Both countries thus decided, based on the relevant provisions of the UN Convention on the Law of the Sea, to establish a JDZ to cover the whole coastal area of overlapping claims within their potential economic exclusive zone in the Gulf of Guinea through a treaty signed on 21 February 2001. The treaty entered into force for both countries in 2003. See J. Tanga Biang "The Joint Development Zone between Nigeria and Sao Tome and Principe: The Case of Joint Development in the Gulf of Guinea – International Law, State Practice and Prospects for Regional Integration", Report of United Nations, The Nippon Foundation Fellowship Programme 2009/10 submitted to the Division for Ocean Affairs and the Law of the Sea Office of Legal Affairs, United Nations, New York.

Nigeria has had a troubled history of misgovernance and systemic corruption. This has underpinned the crisis of development in which the country is immersed, and undermined the country's image in the last two and half decades. The link between the oil/gas sector and corruption in the country has been rightly drawn by Okonjo-Iweala and Osafo-Kwaako (2007:17) when they assert that "a bane for Nigeria's existence since the oil boom of the 1970s has been the reputation for corruption ..." Lubeck, Watts and Lipschutz (2007) estimated that Nigeria lost between US$50 to US$100 billion to corruption and fraud in the oil sector, while a news report by Global Financial Integrity puts the figure for illicit financial outflows from Nigeria between 1970 and 2008 at US$58.5 billion (Kar and Cartwright-Smith 2010).

Referring to the pervasiveness of corruption in public institutions in Nigeria, Okonjo-Iweala and Osafo-Kwaako cite the findings of a 2002 survey: 70 per cent of firms surveyed indicated they paid bribes to obtain trade permits; 83 per cent reportedly paid bribes to access utility services; 65 per cent paid bribes while paying taxes; and an estimated 90 per cent paid bribes to facilitate procurement. In the same survey, 70 per cent acknowledged the need to pay bribes to secure favourable judicial decisions, while 100 per cent shared the widespread view about the diversion of public funds into private use, compared to 78 per cent and 45 per cent of firms in Russia and South Africa respectively. Figure 1 below provides a graphic presentation of this trend.

FIGURE 1. LEAKAGES OF PUBLIC FUNDS IN NIGERIA, RUSSIA AND SOUTH AFRICA

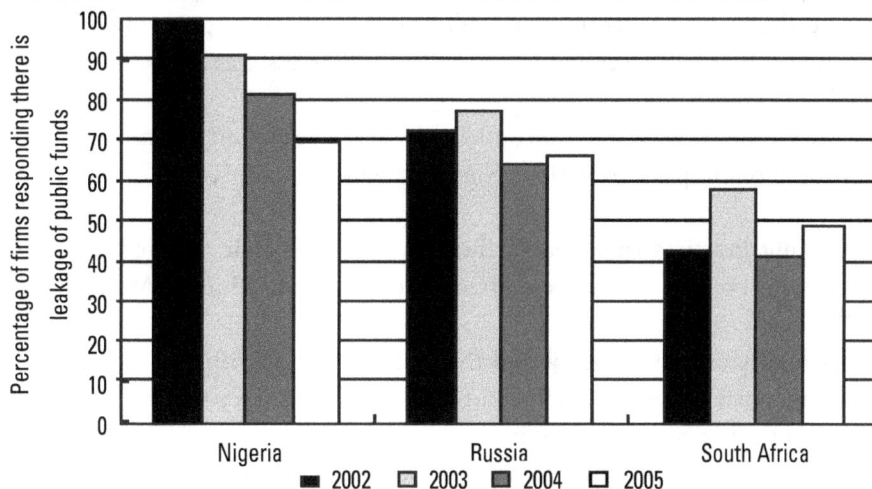

Source: Danny Kaufmann et al. (2005), "Nigeria in Numbers – the Governance Dimensions: A Preliminary and Brief Review of recent trend on governance and corruption", a presentation for the President of Nigeria and his Economic Management Team (Mimeo). This work is also cited in Antoine Heuty, "The Challenges of Managing Oil and Gas Revenues in Nigeria", a presentation to the budget training workshop organised by Revenue Watch Institute for its Nigerian partner organisations, 8–16 June 2009.

Cases of corruption have dotted the Nigerian landscape since colonial times. The African Continental Bank (ACB) case could be said to be the first major celebrated (in the sense of wide publicity and politicisation) case of corruption, in that it involved the then foremost Nigerian nationalist, the leader of the National Council of Nigeria and the Cameroons (NCNC, which later became National Council for Nigerian Citizens) and premier of Eastern Region Dr Nnamdi Azikiwe. In 1957, the colonial government sought to address the issue through an ad hoc panel to look into the charges of corruption against Dr Azikiwe. In 1962, the federal government used a similar approach to address the next high profile case of corruption, involving Chief Obafemi Awolowo and the leadership of the Action Group (AG), the leading political party in Western Nigeria and the main opposition political party in the federal parliament. This time it was the Coker Commission. These investigation panels were ad hoc in character and their legitimacy was questioned, given that they were set up by governments to investigate political opponents. Thus, the reports of the investigative panels often failed to convince supporters of the accused of the fairness of the conclusions reached. Such probes were seen as orchestrated witch hunts intended to tarnish the reputation of the opposition. Several anti-corruption investigative panels were set up during the military era (1966–79, 1984–99). Of note was the Okigbo panel that investigated how the Gulf War oil windfall was utilised by the federal military government under General Ibrahim Babangida (1985–93). The Okigbo report (whose contents are yet to be made fully public) found that over US$12 billion from crude oil proceeds was missing from the national coffers. The Okigbo panel was initiated by the military regime that succeeded the interim government installed by the Babangida government before his departure from power and investigated its predecessor as a way of legitimising its own palace coup.

Over time, various anti-corruption laws/legal frameworks and institutions have been put in place by various Nigerian regimes (Ajayi 2003). These include the:

- Corrupt Practices Investigation Bureau and the Public Complaints Commission, both of which were established under the Murtala Mohammed/ Obasanjo regime;
- Code of Conduct Tribunal and the Code of Conduct Bureau, which were inaugurated under the 1979 Constitution and have since become part of all subsequent Nigerian Constitutions;
- Recovery of Public Property (Special Military Tribunal) Decree No. 2 of 1984; Special Tribunal (Miscellaneous Offences) Decree No. 20 of 1984;
- Public Officers (Protection Against False Accusations) Decree No. 4 of 1984, and the Counterfeit Currency (Special Provisions) Decree No. 2 of 1984;
- Failed Banks Decree of General Abacha regime; and

- Abdulsalami Abubakar regime's Forfeiture of Assets, etc. (Certain Persons) Decree No. 53 of 1999.

It is worth noting that some prominent public officials/individuals were prosecuted for corrupt practices and sentenced to various jail terms. However, most prosecutions took place at the inception of a new military regime and were often part of a ploy to win legitimacy by adopting an anti-corruption stance and probing the immediately preceding administration. Such zeal and political will to pursue corruption cases against former officeholders tended to decline as the succeeding regime adopted the habits of preceding ones.

Various regimes initiated attitudinal change programmes aimed at curbing corruption with varying degrees of success. Examples included the Ethical Revolution of the Shehu Shagari administration, the War Against Indiscipline (WAI) of the General Muhammadu Buhari regime and the Mass Mobilisation for Social and Economic Recovery (MAMSER) introduced by the military President Ibrahim Babangida regime. General Abacha's military regime also introduced the War Against Indiscipline and Corruption (WAIC). With the exception of WAI under the Buhari regime, it is difficult to attribute any positive impact to these attitudinal change programmes, as the demise of the regimes that introduced them generally produced revelations of massive corruption by political leaders.

More recent anti-corruption initiatives and legal frameworks are:
- The Corrupt Practices and other Related Offences Act 2000, with an accompanying Independent Corrupt Practices Commission (ICPC) as the implementing body;
- The Fiscal Responsibility Act and the Public Procurement Act 2007 and the institutional mechanism for due process (i.e., Fiscal Responsibility Commission and Bureau of Public Procurement);
- EFCC; and
- NEITI

The foregoing initiatives and institutions were established in the context of Nigeria's return to democratic rule after August 1999. These anti-corruption programmes were also geared to streamlining processes of doing government business by plugging loopholes that could facilitate corrupt practices, and to avoiding the malfeasance associated with "business as usual". Other initiatives deploy investigative and prosecutorial authority in the fight against corruption, even though the process has attracted allegations that certain individuals were being targeted for prosecution for political reasons. Such criticisms notwithstanding, the novel anti-corruption initiatives associated with the governance and institutional reforms introduced by the Obasanjo administration received a boost with the inception of NEITI.

As a subset of the global EITI, NEITI's mandate is transparency and accountability in the extractive industry, which, given the centrality of the oil and gas industry to the Nigerian political economy, is crucial. The uniqueness of EITI in Nigeria lies in the fact that it provides oversight at the revenue end of the fiscal chain, drawing in various governmental and private business organisations that are involved in revenue-generation within the oil and gas industry, and ensuring a production-revenue match and an inter-agency receipt match. This is no doubt a giant step in the anti-corruption efforts in Nigeria. EITI monitors and charts the flow of revenue from oil, gas and mining companies to host governments, publicises them, on the basis of which citizens can hold their governments accountable (Ikubaje 2006: 55). This puts civil society squarely on the demand side of transparency and accountability spectrum. Indeed, that role is an institutional one within the context of the global principles and criteria of the EITI process developed and adopted at the Lancaster House Conference in June 2003.[3] One of the major innovations of EITI is the collaborative element between host governments and extractive companies built into its operations. Another one is the recognition that if revenues accruing to the state must have a democratic and developmental impact, the public also must be a critical member of this partnership. The Nigerian EITI process, even though connected to the global EITI process, shares some internal peculiarities that make it relatively different from the processes in several other countries.

Institutional framework of the NEITI process

The institutional framework of the NEITI process is anchored on the general principles and criteria of EITI as developed and adopted at the Lancaster House conference. Both at country and global programme implementation levels, these principles and criteria serve as the cornerstone of the EITI process. The EITI principles stress the prudent use of natural resource wealth as an important factor in economic growth and sustainable development; healthy democratic debates and informed choice of realistic options for national development; and improved public financial management and increased standards of transparency and accountability in public life, government operations and in business. It also promotes an enhanced environment for domestic and foreign direct investment, and recognises the contributions of stakeholders (government and its agencies, extractive industries and their associated service companies, multilateral organisations, financial organisations, investor and non-governmental organisations).[4]

Based on the foregoing principles, the implementation of EITI is consistent with criteria such as regular publication of payments by extractive companies to

---

3. The principles and criteria have remained the most concise statement of the beliefs and aims of EITI.
4. See "The EITI Principles and Criteria", available at: http://eiti.org/eiti/principles

governments and the latter's acknowledgement of same in the most accessible and comprehensible manner; reconciliation of all payments and revenues by an independent and credible administrator applying international auditing standards and providing a report about discrepancies, wherever identified, about the exercise; and extending the reconciliation exercise to all companies (including state-owned). The criteria also include providing space for civil society to engage actively in the design, monitoring and evaluation of the process, and a joint stakeholders' financially sustainable workplan with measurable targets. Figure 2 below demonstrates the EITI process.

FIGURE 2. DIAGRAMMATIC VIEW OF THE EITI PROCESS

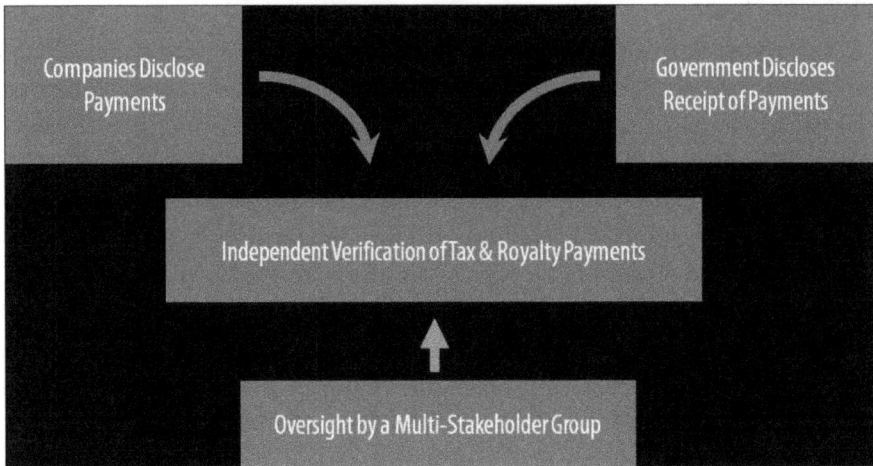

Companies Disclose Payments

Government Discloses Receipt of Payments

Independent Verification of Tax & Royalty Payments

Oversight by a Multi-Stakeholder Group

Source: Extractive Industries Transparency Initiative, *EITI Fact Sheet,* EITI Secretariat, Oslo, 25 November 2010

Any commitment to reconciling company payments and government revenues through a multi-stakeholder approach also directly commits parties to good governance, international credibility and the fight against corruption (EITI 2010: 1). The uniqueness of NEITI lies in the manner in which it attempts to come to grips with corruption by focusing on a dimension of the "resource curse" as it afflicts Nigeria. This curse refers to the general tendency for countries well endowed with natural resources to "perform worse in terms of economic development and good governance than do countries with fewer resources" (Humphreys, Sachs and Stiglitz 2007: 1). One dimension of this affliction relates to the fact that dependence on the extractive industry often leads to neglect of productive sectors such as agriculture and manufacturing. The other relates to the rentier character of the extractive sector in the Nigerian economy, which undermines its productive capacity. As public finance and foreign exchange earnings increasingly rely on rents from the oil and gas industry, there is the widespread feeling that processes of generating revenues from this industry have been overwhelmed by corruption. The task of NEITI is to promote transparency and accountability and confront corruption head on.

Scope of the study

This study is based on information gathered from numerous individuals representing the very diverse stakeholders in the NEITI process. These included in-depth interviews and focus group discussions with civil society organisations, oil and gas companies, officials of the CBN, the EFCC, the Nigeria partner firm of NEITI auditors (S.S. Afemikhe and Co.) and officials of the NEITI secretariat. The study also benefited from a meeting convened by CISLAC, where a civil society position paper was developed and adopted as a memorandum for the PIB public hearing.

This report is divided into five chapters, including this introduction. Chapter two explores the processes and structures for promoting accountability and tranparency in the oil and gas industry in Nigeria, while the third chapter analyses the dynamics of the interactions between various groups in civil society – NGOs, oil and gas workers, the media, and various government agencies and law enforcement agencies, including NEITI. In this regard, the roles of the CBN, EFCC and of course, the NSWG are also examined in the context of the NEITI process. Chapter four focuses on emerging issues in the extractive industry in Nigeria that may not have been covered by the NEITI Act. This includes a brief review of the PIB, currently before the National Assembly and the possible ways it could impact the NEITI process. It is followed by chapter five, consisting of the conclusion and key policy recommendations geared to mainstreaming NEITI as an anti-corruption agency and catalyst for Nigerian development.

2. BUILDING TRANSPARENCY AND ACCOUNTABILITY IN THE OIL INDUSTRY

Matching production with receipts: The audit process

The auditing of the oil industry in Nigeria is perhaps the most notable innovation aimed at policing the extractive sector under NEITI. A comprehensive auditing of oil and gas receipts had never been undertaken or even contemplated until the NEITI-inspired audit of 1999-2004. The extraction of oil and gas in Nigeria is partly based on the Production Sharing Contract (PSC) in which profit is shared after recovery of costs by the contracting parties. These parties are the oil producing companies and the federal government of Nigeria through the NNPC. Apart from the "no-go areas" and virtual black boxes the oil and gas companies have erected to render actual costs of production opaque, the government institutions involved in the oil and gas revenue chain tend to act in isolation from one another, to the point of adversely affecting expected deliverables.[5] Information on revenue was available from different sources, but government agencies apparently in charge of assessing, collecting and ensuring the safekeeping of such revenues hardly coordinate their activities and record keeping. In this context, there was a strong feeling that oil companies could easily short-change government by under-reporting oil production and inflating costs. It was also suspected that revenue leakages at various agency levels could hardly be noticed.

It is within the context of the PSCs that the Nigerian government demanded its unpaid signature bonuses running into billions of dollars since 1999.[6] Figure 3 below gives details of revenue accruals from signature bonuses to Nigeria between 1999 and June 2007.

The NEITI audit process took on the oil and gas industry head on. For the first time an audit was carried out at the various stages of oil extraction, production and export: the wellhead, flow station and export terminal. It was able, on the basis of this approach, to match oil production with volume exported and payments/revenues. It was also able to show the hydrocarbon mass balance, which indicates the total volume produced, what was exported and what was lost to theft or spillage. Before this, the actual amount of oil production was unknown. The data available related only to the volume officially exported, and the quantity of crude oil stolen or lost to spillage was unknown. In effect, the

---

5. This came out clearly in interviews and public discussions by representatives of these institutions.

6. See John-Abba Ogbodo, "Reps ask oil firms to pay N225.45 trillion", *The Guardian* (Lagos), 25 February 2009, (available at: http://speakersoffice.gov.ng/news_feb_24_09_1.htm); Bassey Udo, "FG Demands $231 Million Signature Bonus From Korean Firm for Two Oil Blocks", Daily Independent (Lagos), 3 November 2008 (available at: http://allafrica.com/stories/200811040904.html); "65 per cent of Oil Signature Bonus Remains Unaccounted" (available at: http://www.financialnigeria.com/NEWS/news_item_detail_archive.aspx?item=2746

FIGURE 3. ACCRUALS IN RESPECT OF SIGNATURE BONUS FROM 1999–2007

Years	Oil Blocks	Companies that made payments	Accounts	Amounts
1999	249, 314, 244, 245, 229, 248	Oil and Gas Nig Ltd, Paragon Petroleum Ltd, Totex Nig, Malabu Oil and Gas Ltd, Integrated Petroleum Coy and Zebra Energy	CBN/PTDF US$ Acct with UBA	$26,190,000
2000	214	EXXOMOBIL	PTDF US$ Acct with UBA	$5,000,000
2001	320, 229, 250	Orandi Pet Ltd, Amni, Emerald Energy, Chevron, Shell, Brasoil Service and Petrobras	CBN/PTDF Acct with UBA and CBN/PTDF Reserve Acct	$102,000,000
2002	324, 214, 318, 214, 244,	Petroleo Brasileioro, Chevron, NPDC, Philips Explode, Gesso and NAOC	Fed Govt Independent Rev Acct	$97,000,000
2003	248, 245, 221, 257	Zebra Energy, Ocean Energy, Statoil, Shell TotalFINAElf, Vintage Oil	PTDF Reserve Acc in CBN, Fed Gov Ind Rev Acct and PTDF (DPR Capital Budget Funding)	$102,500,000
2004	OML14 OML88 OML95 OML90 OML55 OML46 OML90 OML13 OML38 OML54 OML30 OML35 OML90 OML11 OML14 OML16 OML11 OPL247 OPL242 OML40 OML56 OPL322 OPL56 OML16 INTEREST	Universal Energy, Goland, Guaranty Sogenal, Del-Sigma, Movido, Excell, Bayelsa Oil, Britamia-U, Network Exp, Platform Petroleum, Eurafric Energy, Independent Energy, Owena Oil, Bicta Energy, Prime Energy, Associated Oil and Gas, Morri, Prime Energy, Damsaki, Frontier Oil, Millennium Oil, Chevron, Ocean Energy, Sahara Energy Field and African Oil and Gas	PTDF (DPR Capital Project) and PTDF Reserve Acct in CBN	$57,570,030
2005	(Names of some blocks are not indicated here) 722, 733, 231, 251, 280, 325, 251, 277, 722, 723 233, 732, 917, 332, 276, 283, 315, 907, 257, 277, 905, 332, 278, 809, 810, 917, 135, 321, 323, 282, 236	ELF, Orient Petroleum, Shell, Energia Ltd, , Midwestern Oil, New Nig Dev Co, Monipulo, NPDC/Refinee Petroplus, Technical Sys/Sterline Global Oil, NPDC/ Ashbert, Amni Inter, Domon Oil Services, Ascon Oil, Boston Energy, Centrica, Petrobras/Statoil, Allen E&P, Conoil, Sterline Global, Gas Trans and Power, BG Exploration/Sahara Oil and Gas, Oando, New Nig Dev Co, VP Energy, Global Energy, NAOC, Equator Exploration, KNOC and Agip	Consolidated Revenue Acct, PTDF Reserve Acct in CBN, CBN Independent Rev Acct, CBN/AGF/FGN Account	$1,062,246,302
2006	471/298, 721, 732, 252, 292, 284, 286, 289, 233, 279, 285, 297, 281, 294, 291	Transnational Corporation, China National Petroleum, INC Natural Resources, BG/ Sahara Energy/Lotus, Clean Water/ NIGDEL United, ONGC MITTAL Energy/ Emo Exploration and Starcrest/Addax	Consolidated Revenue Acct, PTDF Reserve Acct in CBN, Independent Rev Acct, CBN/AGF/FGN Account	$404,629,890
2007	226, 231, 290, 2009 and 2010, 2007, 2005 and 2006, 240, 293, 258, 228	Essar Energy, Monipulo, Conoil, Global Energy, Continental, Sterline Global Oil, Bayelsa Oil Co, Abbey Court/Coscharis, Delgate/Petrodel, Sahara Energy Field	CBN/AGF/FGN Acct, Consolidated Revenue Acct	$228,917,345

Source: Economic Confidential, February 2008. Available at: http://www.economicconfidential.com/febfactdpr.htm

audit provided information on the quantity of oil and gas produced (physically measured); the financial report that indicated the amount of money actually paid for what was produced; and a process audit, which tried to unravel the amount exported against that allocated for domestic use, refining, etc.[7]

The Nigerian EITI process went a step beyond what the EITI recommends. The EITI process essentially focused on the amount of money paid and what was received. The NEITI process (EITI++ as its World Bank version is dubbed) not only focuses on verifying and matching the volume of exports and amounts paid, it also examines the propriety of the process by which the figures were generated and compiled. Thus the NEITI process is more rigorous in certain regards:

- It moves beyond a reconciliation *to audit by* verifying and investigating discrepancies identified;
- It *conducts Value-for-Money audits by* auditing underlying cost structures in major projects;
- It undertakes benchmarking by comparing performance of projects against other domestic projects and international comparables;
- It probes *physical audits by* measuring the physical commodities extracted and associated processes;
- It *examines industry processes to ascertain* licensing, portfolio management, state investments and local content; and
- It moves beyond revenue generation to distribution by ascertaining allocations, distribution, receipt and usage, at the regional and local levels (Revenue Watch Institute 2008:48).

The above makes the EITI process in Nigeria a comprehensive approach, even though the audits are more expensive and time-consuming to produce (Revenue Watch Institute 2008:48). Symbolically, it is intended that the physical and the financial audit have a meeting point, and "shake hands". The scope of the three different kinds of audit conducted by the Hart Group in association with S.S. Afemikhe and Co. were:

- *Financial audit:* Who paid money? How much? To whom?
- *Physical audit:* A mapping of oil and gas produced, refined, exported, lost, etc.
- *Process audit:* Examination of extractive industry processes in licensing, capital expenditure proposals, etc.

The 1999–2004 NEITI audit achieved such successes as reducing the initial wide discrepancies in payments (of taxes, royalties and others, such as gas flare

---

7. Interview with Sam Afemikhe (Lagos), 3 August 2009.

penalties, etc.) by oil companies to the federal government to a narrow margin (0.02 per cent), reconciliation of physical flows up to terminal and account-ability of all crude sales. However, the audits, including the latest one covering 2005, also uncovered discrepancies between CBN's accounts of receipts and payments made by oil and gas companies operating in Nigeria. Perhaps because of its broad scope and depth, the 1999–2004 audit revealed shocking systemic weaknesses in the key processes examined. These weaknesses were classified in the three categories of the audit process:

Financial audit:
- Weak coordination/interface among relevant government agencies
- Poor data-keeping system, resulting in revenue fluctuations – e.g., CBN's re-cord showing lower figures for payments to the government's account than what oil companies reportedly paid.

Physical audit:
- Systemic loss of crude oil between wellhead, flow station and export terminals
- Flow rate at night showing lower records than during the day, suggesting theft
- Poor precision metering infringing on gross production volumes
- Absence of standardised industry procedures for calculating royalty liability
- Myriad other issues associated with the handling of imports of petroleum products.

Process audit:
- Abuse of discretionary powers (based on the Petroleum Act of 1969) of the minister of petroleum resources to allocate oil blocs
- Poor implementation of Local Content Vehicles (LCVs) in the 2005 bid
- Arbitrary increases in the use of strategic downstream investment considera-tions that are tied to upstream oil blocs
- Inconsistencies in the procedures for awarding petroleum contracts/policies.

Although these differences were considered minimal, being less than 2 per cent as against the conventional margin of error for auditors, some civil society or-ganisations such as PWYP-Nigeria and Zero-Corruption Coalition used the audit findings as a basis for engaging the NEITI secretariat. These discrepancies were explained away by CBN as the result of differences in reporting templates. For example, while CBN captures revenue on a company by company basis, oil companies do their reporting on the basis of oilfields. The physical aspect of that audit discovered discrepancies between volumes of crude recorded at oil well-heads and at flow stations. This was again taken up by the civil society organisations with NEITI and, through it, with oil and gas companies. The explanation the oil companies gave is that wellhead output generally contains a great deal of water and natural gas, which would have been separated from the crude at the flow stations. The search for a final solution lay behind the request by President Yar'Adua government for assistance from the Norwegian government. However, there are allegations that the initiative has been largely sabotaged by NNPC.

The 2005 audit report released in September 2009 indicated that there was a difference of 1.05million barrels between the physical amount of oil production and the financial returns made by the Nigerian oil industry. The same 2005 audit also pitched the auditors against officials of the DPR and FIRS over a US$524 million underpayment in respect of the budgetary reference to oil prices for that year. The EITI process has begun pinpointing the major revenue leakages. Agencies and companies have come to learn through the findings of the audit that whatever was "messed up" has to be "cleaned up" through refunds or restitution. This situation indicates a major impediment to corruption, especially as NEITI has continued to draw attention to the non-remittance of N345billion (US$2.3billion), this being the difference identified in payments made to the federal government in NEITI's 2005 oil and gas audit (Esiedesa 2010).

However, its anti-corruption value depends on what NEITI does with the information available to it, and the impact of its disclosures on governance and accountability in the larger society. In fact, this is a major challenge for the implementation of the EITI process. A total of 25 EITI reports have so far been produced by implementing countries from 2003 to September 2010, all varying in terms of sectors covered, level of data aggregation/disaggregation and regularity/currency of reporting cycles. The major challenge regarding many of these reports, including those for Nigeria, is what to do with the mass of data and information generated by the EITI process.

One of the major successes of the EITI audit process lies in its capacity to open up access to information. According to the auditors, all relevant agencies have been generally cooperative in providing information for the audit. The state power behind the EITI process in Nigeria facilitated accessing the information, but the international standing (moral suasion) of the global EITI process was also critical in getting foreign oil company cooperation in providing information. The fact that NEITI auditors are present is seen as a huge deterrent to corruption. The age of producing figures that are not subject to verification or reconciliation is gone. In this regard, the EITI process can be seen to have considerably widened the anti-corruption space in Nigeria by promoting transparency and accountability in extractive resource governance. However, this reading needs to be nuanced in terms of the potential of the process rather than the actuality. Actualisation will depend on the activation of the NEITI enforcement machinery. Civil society organisations are critical in this regard.

In this connection, two major shortcomings of civil society must be noted. The first relates to the limited technical capacity to understand the production process and sieve through complicated financial audits. The second relates to the possibility that the relative ease of access to information that underpinned the first two audits may not be guaranteed under subsequent audits. This is due to changes – or perceived changes – in legal regimes. While the first audits were largely done in a climate of strong presidential backing for the process, but without an enabling law, a NEITI Act has since come into being in 2007. However, the act is seen as removing NEITI's teeth, since information prejudicial to the interests of oil companies or government cannot be used by NEITI to institute legal action against them. Specifically, Section 3 (d and e) of the NEITI Act 2007 states that for the purpose of realising its objectives under the Act, NEITI shall perform such functions including:

> (d) obtain, as may be deemed necessary, from any extractive industry company an accurate record of the cost of production and volume of sale of oil, gas and other minerals extracted by the company at any period, *provided that such information shall not be used in any manner prejudicial to the contractual obligation or proprietary interests of the extractive industry company* (e) request from any company in the extractive industry, or from any relevant organs of the Federal, State or Local Government, an accurate account of money paid by, and received from the company at any period, as revenue accruing to the Federal Government from such company for that period; *provided that such information shall not be used in a manner prejudicial to contractual obligations or proprietary interests of the extractive industry company or sovereign obligation of Government.*[8]

In spite of this, there is a feeling that civil society can fill the void by insisting on policies or sanctions against observed financial discrepancies.

Before the advent of the EITI process in Nigeria, various government departments involved in revenue generation from the extractive industry were literally not on talking terms. This was not because of any defined animosity, but was more the product of the absence of a coordinating institution or process to facilitate interaction. The need for such interaction was not even apparent. EITI has defined and provided the platform for a thorough audit process. This process has also exposed many of the flaws in the extractive industry in Nigeria. For example, to get a list of industry operators from a single source was just not possible. Several firms have mining licences, but neither NNPC nor the petroleum ministry had a list of these operators in the industry. The audit firm of S.S. Afemikhe and Co. had to piece this information together from diverse sources over nine months using the internet, newspapers and other informal or non-official records. Normally, information of this nature ought to be readily available, but

---

8. Emphasis added by author.

it was not. Yet this kind of information is vital in carrying out physical, fiscal and process audits and in determining the hydrocarbon mass balance. The determination of the latter, perhaps for the first time, gave an accurate clue as to the quantity of crude produced in Nigeria and its cash value. In Nigeria, a mass hydrocarbon balance had never been calculated as the physical audit revealed the inability to undertake measurements at the wellhead. This difficulty led to the practice of determining the volume at the point of export, as this became by default the most effective point for accurately measuring the volumes (interview, Sam Afemikhe, 2003).

It is not clear whether this lack of information as to industry operators was deliberate or the result of plain incompetence among the government agencies that ought to possess such data. However, it does point to the fact that the operations of the oil industry were "personalised" and were largely defined by and connected to political power in Nigeria. It also showed how the oil-power nexus rendered oil-revenue governance opaque, an opaqueness that virtually concealed the illegal diversion of massive revenues by highly placed government officials and their hangers-on. The petroleum minister virtually personalised the licensing of oil blocs. There are reports that such blocs were arbitrarily allocated on a patronage basis to politicians, traditional rulers, top military officers, cronies, etc. These individuals in most cases turned around and sold such oil blocs (licences) to real producers at huge profits. A vivid example was the revelation by retired Lt-General T.Y. Danjuma of how he made US$500 million from an oil bloc he was allocated, which he later sold for US$1 billion. Nigeria's former chief of army staff (1976–79) and minister of defence (1999–2003) told his audience at the public launch of the T.Y. Danjuma Foundation on 10 February 2010 that 12 years earlier he had been allocated an offshore oil bloc by the regime of late General Sani Abacha, adding that it took him ten years before his company struck oil in the bloc, which he then decided to sell for the US$1 billion.

Licences sometimes changed hands three or more times to the point that it became difficult to track actual ownership at any one time. Precisely for this reason transparency in the licensing of oil and gas operators should be vigorously canvassed by NEITI as a prerequisite for effective monitoring of oil and gas revenues. The PIB will hopefully address this issue.

The determination of the hydrocarbon mass balance has also uncovered practices of the oil operators that may have adversely affected oil-derived revenues. Before the audit, producing oil firms "worked from the answer to the question". Basically, it was thus possible to simply attach volume produced to the revenue declared. This sum might and could be very inaccurate. In any case, this methodology lacked transparency, and gave wide and largely unregulated operational scope to oil companies. While the commitment to the EITI process has facilitated two audits of the Nigerian oil industry, no audit has been carried out yet under the NEITI Act 2007. As noted earlier, the main problem in carrying out an audit under the act is Section 3(d and e), which may lead oil companies to withhold information from the auditors and which prevents such information, if given, from being used to bring an erring oil company to book. This provision is a very serious impediment to using sanctions to deter corrupt behaviour. Consequently, CSOs took the opportunity during the public hearings on the PIB to advocate that its robust information disclosure provisions be extended to the NEITI Act by simply adding that the provisions of the latter shall hold only to the extent they are consistent with the former (assuming that the particular section of the bill will pass as submitted to the National Assembly). They argued in their submission on the PIB that it should address the confidentiality clauses in the NEITI Act 2007 as follows:

> The Institutions and National Oil Company shall be bound by the principles of the Nigeria Extractive Industries and Transparency Initiative Act of 2007 and where the confidentiality clause of NEITI Act conflicts with Section 273 of the Petroleum Industry Act, the Petroleum Industry law shall take precedence.[9]

The proposal is aimed at eliminating the opaqueness and secrecy identified as the key ingredients of corruption in the oil and gas industry, particularly as it relates to royalties, fees, bonuses of whatever sort and taxes. Civil society further requested that the pending Freedom of Information Bill (FOI) before the National Assembly be given the passage it deserves to strengthen the PIB.[10]

Finally, there is the issue of remediation, which formed part of the recommendations of the 1999-2004 audit. The remediation plan covers five key areas – developing a revenue-flow interface among government agencies, improving Nigeria's oil and gas metering infrastructure, developing a uniform approach to cost determination, building human and physical capacities of critical government agencies, and improving overall governance of the oil and gas sector – and was drawn up by the IMTT inaugurated by the Obasanjo government to address the lapses identified in the audit report.

The remediation that has actually flowed from these recommendations is what the oil companies took upon themselves to correct. At the level of governmental agencies (DPR and FIRS), not much has been done. Intergovernmental meetings of these bodies, including CBN and NNPC, have, however, commenced. Although the various public agencies identified have made many promises about effecting the necessary changes, minimal remediation is the

---

9. See "A Memorandum on the Petroleum Industry Bill 2009 Submitted to the House of Representatives" by Civil Society Working Group on Extractive Revenue Transparency, Accountability and Good Governance in Nigeria, p.2; "A Memorandum on the Petroleum Industry Bill (PIB) 2009" submitted by the Niger Delta Budget Monitoring Group to the Senate and House of Representatives, p.4.

10. Ibid.

order of the day and the agencies "are still as they are", to quote a reliable source directly involved in this process.[11] Oil quantity, for instance, is still determined by the "dipstick". However, the advent of NEITI has at least shown that things could be done differently and more efficiently. NEITI has initiated a study on the metering system with a view to effecting a modality to accurately gauge the volumes of oil produced and exported. The contract, awarded to Telemetri Nigeria Limited and paid for with financial assistance from the UK's Department for International Development, is broadly divided into upstream and downstream sectors. The former seeks to "develop a strategy for metering that is capable of providing government with the necessary information to quantify production and pipeline losses with a view to improving management of resources, environmental impacts, federation revenue and related matters". Meanwhile, the latter seeks to "improve metering and general management accountability for pipeline losses by enhancing available inputs for efficient and effective downstream hydrocarbon mass balance" (Orogun 2009).

The foregoing is a sequel to an earlier attempt to elicit support from the Norwegian government on this critical aspect of remediation. Several government agencies were required to submit proposals on ways to fix the human and infrastructural gaps in the oil industry. There was also a facilitated discussion with the German government and a German software company – SAP – for possible support for the Revenue-Flow Interface Project, while the Commercial and Marketing Department (COMD) of NNPC was, in an effort to address the absence of uniform pricing for the different grades of Nigerian oil, mandated to determine such prices and advise the FIRS and companies accordingly. Oil companies that do not comply are to be prevented from pumping crude. Another issue is the manual form for recording information on crude sales. The auditors have taken on NNPC on this issue, but adopting more technologically savvy methods has not been embraced with any urgency by the national oil corporation. As for DPR, it has largely failed to engage the auditors on remediation. The FIRS readily attracts assistance from donors, but insists it must be in sole control of the deployment of such funds, based on its own determinations and priorities.

The PIB appears to be one commendable effort to confront lapses revealed by the audits. That in itself derives from IMTT's resolve that improving governance of the oil and gas sector will require renewal of the laws regulating it.

---

11. This was the situation in October 2009, when the fieldwork for this study was conducted. A few actions have been taken since to demonstrate that some corrective measures are being gradually instituted.

Enforcement and justiciability of the NEITI Act 2007

There is great scepticism within CSOs about the prospects of enforcing the NEITI Act. The act clearly enumerates violations that can be prosecuted under its terms. These are reasonably exhaustive, as evidenced in Section 16 of the Act. The stipulated sanctions are also stiff, given that individuals as well as corporate organisations may be held criminally liable. Apart from restitution of revenue lost, sanctions could include jail terms for individuals and fines for both individuals and corporate bodies. The latter could also have their licences suspended or revoked.[12] The reservations by elements of civil society about the enforcement of these provisions lies in the manner the act undermines itself.

First, the Act stipulates that liability cannot be placed on an individual if s/he can show the offence was committed without his consent or connivance, or that s/he took all necessary actions to ensure that the crime was not committed.[13] Second, as already noted, while the Act stipulates that extractive industries must provide NEITI with accurate records of production and volumes of sale if so requested, it blunts the potential deterrent effect of this provision by insisting that any information so obtained "... *shall not be used in any manner prejudicial to the contractual obligation or proprietary interests of the extractive industry company".*[14] Similarly, the NEITI can request accurate information from extractive companies on payments made to any level of government or ask any of these levels for accurate information on their receipts from any extractive industry. Again, this provision has a proviso: *such information shall not be used in a manner prejudicial to contractual obligations or proprietary interests of the extractive industry company or sovereign obligations of government.*[15]

Quite a few civil society activists hold that these provisions impair the prospects of successful prosecution for infringements of the NEITI Act, except where government intervenes by imposing sanctions on offenders. The original version of the bill did not contain these provisions. However, strong lobbying by oil and gas companies found sympathetic ears in the National Assembly. In terms of "contractual obligations", "proprietary interests" and "sovereign obligations", it is highly possible that the flow of revenues or production volumes can be conveniently hidden. In short, this information may be given on request to the NEITI secretariat, which may be obliged to simply file it away without making it public. Additionally, it may also be impossible to prosecute anyone for infringement on grounds that such action may be prejudicial to state or corporate

---

12. Section 16 (4).
13. Section 16 (4) a and b and (5) a and b.
14. Section 3 (d).
15. Section 3 (e).

interests.[16] Indeed, the act by this clause has been completely watered down as a result of massive lobbying by extractive industries.

The partnership between state and oil companies is reflective of the character of the Nigerian state, which is largely parasitic and clearly unable to resist the external pressures from IOCs. The parasitic character of the Nigeria state is rooted in rentier linkages with oil receipts[17] and reflected in the Joint Venture Agreements (JVAs) between the Nigerian state, through the NNPC, and the six biggest oil companies operating the Niger Delta. Figure 4 below shows that Nigeria owns 55 per cent equity shares in Shell Petroleum Development Company (SPDC), 50 per cent in Mobil Producing Nigeria and 60 per cent each in Chevron Nigeria, Nigeria Agip Oil, Elf and Texaco Overseas (Nigeria) Petroleum, all of which account for 93.9 per cent of total oil production in Nigeria. Figures 4 and 5 below also show Nigeria's equity shares in major oil multinationals and the contractual flow of equity crude through the NNPC.

FIGURE 4. NIGERIA'S EQUITY SHARES IN LEADING OIL MULTINATIONALS OPERATING IN THE COUNTRY

S/N	Oil Company	Shareholders/Share Equity	Operator	Share of National Production
1.	Shell Petroleum Development Company (SPDC)	**NNPC – 55%** Shell – 30% Elf – 10% Agip – 5%	Shell	42.0%
2.	Mobil Producing Nigeria	**NNPC – 60%** Mobil – 40%	Mobil	21.0%
3.	Chevron Nigeria	**NNPC – 60%** Chevron – 40%	Chevron	19.0%
4.	Nigeria Agip Oil	**NNPC – 60%** Agip – 40%	Agip	7.5%
5.	Elf Petroleum Nigeria	**NNPC – 60%** Elf – 40%	Elf	2.6%
6.	Texaco Overseas (Nigeria) Petroleum	**NNPC – 60%** Texaco – 20% Chevron – 20%	Texaco	1.7%
	TOTAL			**93.8%**

Source: Compiled from Festus Iyayi (2000), "Oil Corporations and the Politics of Community Relations in Oil Producing Communities", in Committee for the Defence of Human Rights, *Boiling Point: A CDHR Publication on the Crisis in the Oil Producing Communities in Nigeria*, Lagos: Committee for the Defence of Human Rights, pp. 155–6; NEITI, *Nigeria Extractive Industries Transparency Initiative: Audit of the Period 1999–2004 (Popular Version)*, NEITI Secretariat, Abuja, n.d., p.7.

---

16. An ongoing case in Uganda involving a whistleblower who leaked a contract signed by the Uganda government to the detriment of the country's long term interest is apt.

17. "A rentier state, according to Omeje, is a state reliant not on the surplus production of the domestic economy or population but externally generated revenue or rents, usually derived from an extractive industry such as oil." See Kenneth Omeje (2005), "Oil Conflict in Nigeria: Contending Issues and Perspectives of the Local Delta People", *New Political Economy*, Vol. 10, No. 3, September, pp.321-34.

FIGURE 5. FLOW OF EQUITY CRUDE THROUGH NNPC

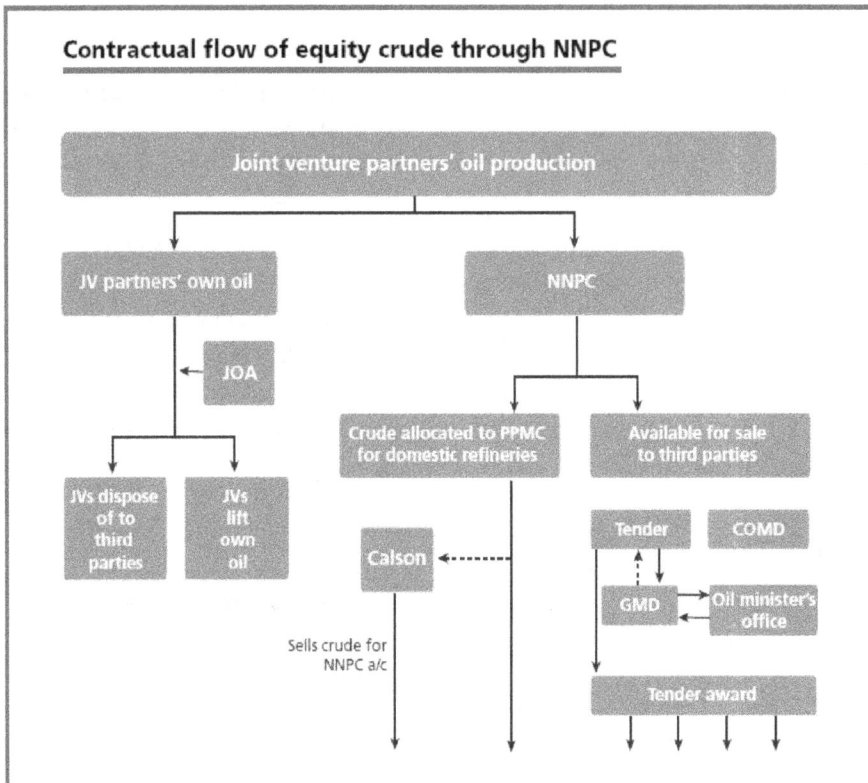

Contractual flow of equity crude through NNPC

Source: Soruce: NEITI, *Nigeria Extractive Industries Transparency Initiative: Audit of the Period 1999-2004 (Popular Version)*, NEITI Secretariat, Abuja, n.d., p.25.

Rather than stand its ground on the new direction for its oil and gas industry as conceived in the PIB, the Nigerian state has allowed its dependence on oil revenues, foreign oil technology and the vagaries of a volatile global oil market to undermine its capacity to resist pressures from IOCs.

In recent months, the presidency and the National Assembly have come under severe attack from civil society and the two umbrella unions of oil workers, which have at separate points accused them of amending the PIB to satisfy the interests of IOCs to the detriment of national interests.[18] In a joint statement signed by Achese Igwe and Babatunde Ogun, respectively the presidents of the Nigeria Union of Petroleum and Natural Gas Workers (NUPENG) and the Petroleum and Natural Gas Senior Staff Association of Nigeria (PENGASAN), it is alleged that:

18. See Chinyere Fred-Adegbulugbe, "Oil workers threaten showdown with FG over PIB", The Punch (Lagos) Monday 27 September 2010, p.17. Available at: http://www.punchng.com/Articl.aspx?theartic=Art201009278373448

The PIB has been drastically amended to essentially favour the interest of the international oil corporations against the Nigerians' quest, and yearnings for the optimisation of our hydrocarbon resources across the upstream, midstream and downstream oil and gas activities through conscious and affirmative policy with measurable milestones. According to reports, 56 changes were made due to the comments made by Oil Producers' Trade Section of the Lagos Chamber of Commerce and Industries. Thirty-six changes were made in response to internal government agencies. Sixty-six changes were made in response to other stake-holders. Some changes were made to reflect indigenous participants' comments. Additional changes made due to other external bodies. Both unions noted that PIB had undergone discrete and selective legislative processes leading to contentious interventions that have caused fundamental reviews of the original draft and the inputs from public hearing while keeping same off-the-shelves and from the website to forestall transparency and easy access.[19]

Both NUPENG and PENGASAN also decried their exclusion from the various legislative and review processes of the Bill, adding:

> … while oil workers, who are the primary operators in the implementation process are consciously excluded in the legislative and review processes, several concessions and compromises have been made at the behest of powers that be, at the dictates of institutions and the privileged, and at the whims and caprices of the barons that can pay the piper.

An Abuja-based newspaper confirmed the foregoing allegation when it reported that a lawmaker who is a member of the three committees in the senate handling the PIB informed it that they were placed under intense pressure from the presidency to accommodate some of the demands of the oil majors. The paper quoted the lawmaker as saying:

> Our intention was to pass the bill as sent to us by the late President Umaru Musa Yar'Adua but these companies put us under intense pressure, they even got the American government to intervene on their behalf. Shortly after his return from the United States early this year when he was acting, President Jonathan requested that the provisions of the bill be reviewed after which he asked the leadership of the two chambers to look at the issue of tax and reduce it to allow for "investment" in the sector[20]

While the provisions for transparency and accountability are still very much alive in the current act, their cohabitation with multiplicities of alibis and exemptions may weaken its ability to seriously challenge corruption in the extractive industry. It is for these reasons that many civil society activists feel strongly

---

19. Ibid.
20. Turaki A. Hassan, "PIB: N/Assembly caves in to oil majors: the Jonathan connection", *Daily Trust* (Abuja), Friday, 8 October 2010.

that the NEITI Act should be sent back to the National Assembly for amendment.

This legislative watering down is further worsened by the fact that NEITI itself does not seem to have the powers of prosecution enjoyed by the EFCC. Indeed, it is expected to hand any matter requiring prosecution to the EFCC, the anti-corruption agency. A few civil society activists hold that the act should locate the powers to prosecute in-house with NEITI, the specialist anti-corruption agency on extractive industry-derived revenues (oil and gas, in particular), which are the main source of foreign exchange and government revenue. This will enhance its ability to deter, and also ensure that its specific focus is not lost within the coils of an already overburdened EFCC.

The current arrangement may not be the best, but it obliges the NEITI secretariat to establish links with other public anti-corruption agencies so as to harmonise procedures and establish operational links, and if necessary conduct joint training. EFCC claims to have formally written on a number of occasions to the NEITI secretariat to attend its anti-corruption programmes, "but they have never responded", claimed an EFCC official. On the other hand, no petition to EFCC has ever emanated from the NEITI secretariat. Yet the extractive industry, especially the oil and gas industry, is, according to EFCC, full of shady deals. There is a suspicion that if no petition, even for investigation, has emerged from the NEITI secretariat, it may be because the secretariat is either not serious in its anti-corruption work or just unwilling to share information. Whatever the underlying reasons, there are hardly any synergies between NEITI and other national anti-corruption agencies.

NEITI bureaucracy and the EITI process

Nothing has prompted public concerns about NEITI since its inception more than the unexpected whistleblowing concerning corruption, cronyism and fraud in its secretariat in 2010. The situation was not only deeply embarrassing and put a burden of explanation on the secretariat, it also, and for the first time, raised questions about the ascribed status of NEITI as "the conscience of the nation in the realm of transparency, accountability and zero tolerance for corruption in the extractive industry".[21]

Two incidents, both intrinsically linked, gave rise to the corruption allegations in NEITI. The first had to do with the 2009 Civil Society Training programmes scheduled to be held in Lagos and Kaduna from 8–13 and 22–28 November 2009 respectively. In this connection, issues of over-invoicing and

---

21. Weneso Orogun had described NEITI as consolidating its status as "conscience of the nation" through its award of meter infrastructure study to Telemetri Nigeria Ltd. See Weneso Orogun, "NEITI Awards Meter Infrastructure Study", *ThisDay* (Lagos), Thursday, 27 August 2009.

payments of N15 million (US$100,000) to hotels[22] without due process were referred to the NEITI board for investigation. The second was the subject of a petition to President Goodluck Jonathan through the secretary to the government of the federation by Stan Rerri, NEITI's director of services, who was sacked over his complicity in the over-invoicing and payments to hotels in Lagos and Kaduna.

On the first issue, the NEITI board set up a committee comprising Leke Alder (chair), Peter Esele, M.I. Yahaya, Mohammed Dikwa (representing Ibrahim DanKwambo, the accountant-general of the federation) and Shehu Sani. The committee's terms of reference were to:

1. Investigate the circumstances that led to the disbursement of funds,
2. Ascertain the exact amount paid,
3. Identify all concerned in the process,
4. Identify culprits and recommend appropriate disciplinary measures,
5. Make other recommendations, and
6. Examine the system of financial management and operation at the time of disbursement.

The committee, which met on 19 May and 2 June 2010, admitted verbal statements and written/print documents from Uche Igwe (civil society liaison officer), Tony Onyekweli (procurement officer), Sukanmi Adeoti (accountant), Stan Rerri (director of services) and Mallam Haruna Sa'eed (executive secretary) as evidence.

All of the committee's recommendations were accepted by the NEITI board at an emergency meeting of 22 July 2010. The recommendations are as follows:

(a) Overhaul of the administration of NEITI Secretariat for efficiency and effectiveness
(b) Review of the Checks and Balances if the Procurement and Contract systems in NEITI
(c) Strict adherence to civil service rules and regulations especially the Procurement Act
(d) All efforts must be made to recover the money in question (including the difference on the inflated tariff on the event that was earlier held in the same hotel), and such must be remitted to the coffers of NEITI.
(e) The following are recommended for disciplinary action for their various roles:
(i) The Executive Secretary [Mallam Haruna Sa'eed]must get a warning from the Board for abdication of responsibility and ineffectual leadership
(ii) Director of Services [Stan Rerri] to be relieved of his responsibility for his roles, action and inaction on this saga

---

22. The hotels were Homegate Resorts and Aso Motel in Lagos and Kaduna respectively.

> (iii) Accountant to be relieved of his responsibility for the roles, action and inaction in the course of these events
>
> (iv) The Procurement Officer to be relieved of his responsibility for his role
>
> (f) That the matter be referred to appropriate security agency.[23]

While it remains unclear if the matter was actually referred to the appropriate security agency as recommended, it is known that the officers recommended for disciplinary action, particularly the three officers identified for disengagement from service, had their appointments terminated.

The second allegation of corruption in NEITI is contained in a widely circulated petition to President Goodluck Jonathan[24] dated 10 August 2010. In the petition, Mr Rerri alleged several wrongdoings within NEITI, including:

- Double dipping in the payment of salaries to the former executive secretary spanning 12 months,
- Collection of honorarium/payment and sitting allowances by the former executive secretary, as well as payment of honoraria to NEITI board members who attend training programmes in Nigeria – over and above their statutory entitlements,
- Collection of allowances by the former executive secretary for sitting on board's sub-committee meetings held at the NEITI office,
- Unregulated pilfering of petty cash account (as advances) in excess of total annual allowances by the former executive secretary, and
- Drawing amounts ranging from US$1,000 to US$3,000 as contingency for foreign travels in contravention of statutory entitlements.

The petition also alleged attempts to undermine President Jonathan's appointed replacement of Mallam Haruna Sa'eed as executive secretary, Mrs Zainab Ahmed, and manipulation of the recruitment process in NEITI to favour cer-

---

23. See Assisi Asobie, "Re-Petition: Fraud at NEITI: An Attempt to Silence the Whistle Blower" (Ref: NEITI/ADM/079/Vol.2/242), a 14-page letter to the Secretary to the Government of the Federation, Alhaji Mahmud Yayale Ahmed, being a response to Stan Rerri's petition to President Goodluck Jonathan, 19 August.

24. The petition which was copied to Alhaji Namadi Sambo (Vice President), Senator Smart Adeyemi (Senate Committee Chairman on Federal Character Commission and Governmental Affairs), Senator Lee Maeba (Senate Committee Chairman on Petroleum – Upstream), Honourable Samson Osagie (House of Representative Committee Chairman on Special Duties), Professor Assisi Asobie (NEITI Board Chairman) and Mrs. Zainab Ahmed (the new NEITI Executive Secretary) was further widely circulated within and outside Nigeria. The media also publicised it. See Kunle Aderinokun, "Crisis Rocks NEITI Over Allegation of Corruption", *ThisDay* (Lagos), Tuesday, 24 August 2010; Bassey Udo, "Corruption Allegation dogs Nigeria's Extractive Industry Monitors", *Next* (Lagos), Saturday, 28 August 2010; Bassey Udo, "Transparency Agency Commences Self-cleansing" Next (Lagos), Sunday, 17 October 2010. Also available at: http://234next.com/csp/cms/sites/Next/Money/5630854-147/transparency_agency_commences_self-cleansing_.csp

tain candidates. Rerri also alleged attempts by the NEITI board chairman to arrogate executive powers to himself, whereas he is a non-executive chairman.

He challenged the termination of his appointment by the NEITI board, arguing that his opposition to violation of due process in NEITI is what put him on a collision course with the former executive secretary, who had earlier threatened "to deal with me".[25] Rerri, perhaps in a bid to win the sympathy of the public, noted that:

> ... as the pioneer staff at the inception of NEITI in 2004, I worked with Mrs. Obiageli Ezekwesili, now Vice President for Africa at the World Bank, to set up and run an Agency of government that was the envy of every resource-rich country in the developing world. In 2007, Nigeria was the world leader in extractive resource transparency. Today, NEITI is about to be ejected from the international EITI body it helped to set up.[26]

In conclusion, Rerri requested President Jonathan to, among other things:
- Empower the new executive secretary "to ensure that Nigeria is not ejected from the Global EITI", as this may affect the country's economic and oil transparency rating
- Assure the EITI secretariat in Oslo of Nigeria's commitment to extractive resource transparency
- Investigate allegations of fraud against the former executive secretary, Haruna Sa'eed, and stop the NEITI board chairman from being a signatory to NEITI accounts or from operating in an executive capacity
- Suspend attempts to replace him, pending the determination of his employment status as substantive director of services overseeing administration and finance,
- Institute an audit into ongoing recruitment process in NEITI, and
- Direct that all NEITI investigation committee reports reflect thoroughness and transparency rather than silence whistle blowers.[27]

For current purposes it is necessary only to analyse those aspects of the petition relating to alleged corruption and the impact of the internal wrangling in NEITI on its statutory role.

The termination of Rerri's appointment by the NEITI board establishes the link between the first and the second incidents of alleged corruption, cronyism and fraud in the organisation. We are not privy to the letter terminating Rerri's

25. Stan Rerri (2010), "Petition: Fraud at NEITI: An Attempt to Silence the Whistle Blower" (Reference: NENTI/PETITIONS/DS/01), being a petition to the President of the Federal Republic of Nigeria through the Secretary of the Government of the Federation dated 10 August.
26. Ibid.
27. Ibid.

appointment, but it is clear from the NEITI board chair's letter to the SGF in response to Rerri's petition, that the latter was relieved of his duties at NEITI on two grounds:

- report and recommendations of the Leke Alder committee that investigated the payments to hotels for NEITI 2009 Civil Society Training,[28] and
- "… his failure to apply for a new position commensurate with his qualification, following the reorganisation of the NEITI Secretariat" – an entirely separate administrative/human resource matter which predates the first ground (2008).[29]

Both reasons are seemingly mutually complementary, if the statement by the NEITI board chairman that Rerri had earlier been earmarked for disengagement on 30 June 2010 (a culmination of issues relating to the work of the human resources consultants in 2008) is anything to go by. That connectedness may be located in the mandate of the Alder committee that investigated the payments to the hotels in 2009, which included: "identify[ing] culprits and recommend[ing] appropriate disciplinary measures, and make other recommendations". Thus, while Rerri may have been originally earmarked for disengagement effective 30 June 2010, his indictment by the Leke Alder committee only brought forward his date of exit. This is the only way to understand the board chairman's statement that:

> Mr. Rerri is no longer staff of NEITI as he had already been informed since June 4 that his services were no longer required, because of his failure to apply for a new position commensurate with his qualification, following the reorganization of the NEITI Secretariat.[30]

Under such circumstances, it should be expected that Rerri, whether in the spirit of the Nigerian project or from a sense that "he that is down need fear no fall",[31] would go public with information on how NEITI "has become a cesspit of corruption, cronyism and fraud". Whatever the motives, the fact remains that the last has not been heard of the matter. Already the NEITI board chairman,

---

28. Ibid, p.2.
29. Ibid , p.13. It is the culmination of an evaluation exercise conducted by human resource consultants on the structure of the NEITI secretariat, staff and their positions and new terms of reference for roles, as well as design of suitable recruitment guidelines and management system. At the conclusion of the exercise, it was alleged that Rerri choose to redesignate himself as "Director of Administration and Finance" without recourse to the sole powers of the NEITI Board on this or recognition of the submission contained in the December 2008 report of the HR consultants that only a Chartered Accountant certification is "a perquisite for this position", with a knowledge and experience in 'Advanced Financial Management' (See ibid, p.3).
30. Ibid, p.13.
31. It is unlikely anybody from either side of the argument could win a clear victory.

Prof. Assisi Asobie, has responded (with supporting documents) to the query by the SGF on the issues raised in Rerri's petition to President Jonathan. This response has also slipped into some hands, including those in the media.[32] A cursory review of the response with specific reference to the allegations against the board chairman has clarified some knotty issues, including reaffirming and rekindling Nigeria's public knowledge of and trust in the integrity of his person and of many others on the board. However, as brilliant and articulate as the response appears, the overbearing actions by the board chairman to protect his hard-earned integrity have been construed within NEITI as self-arrogation of full-time executive powers, which his mandate as a non-executive chairman does not include. This is the only way to interpret the rather rancorous relationship between him and former Director of Services Stan Rerri, on one hand, and former Executive Secretary Mallam Haruna Sa'eed, on the other.

However, while Nigerians and the rest of the world await the outcome of the matter now before the SGF and the presidency, the fact remains that Rerri's "whistle blowing", whether true or false, has inflicted some damage on NEITI's local and international reputation. Dauda Garuba, the Revenue Watch Institute's Nigeria programme coordinator, aptly noted at a public forum:

> I have been engaged with EITI issues with my New York office in the past two weeks ... At every point, the development in the NEITI Secretariat keeps recurring. As a global initiative that Nigeria convincingly led since inception, it is a pity to know of the allegations of corruption within NEITI Secretariat. This is particularly intriguing because it touches the very reasons for which NEITI was set up.[33]

Nigerians are aware of the moral burden they face concerning the stereotyping of citizens of the country as corrupt. In fact, this negative image is what the whole concept of the *Face of Africa* project and its succeeding *Re-branding* project were meant to correct. The odium arising from allegations of corruption in the NEITI secretariat in 2010 was the very last thing Nigerians and the international community needed. If anything, they led to the reinterrogation of the acclaimed notion that NEITI could be the new face and conscience of the nation in the realm of transparency, accountability and zero tolerance for corruption in the extractive industries sector.

---

32. See Bassey Udo, "Transparency Agency Commences Self-cleansing" Next (Lagos), Sunday, 17 October 2010. Available at: http://234next.com/csp/cms/sites/Next/Money/5630854-147/transparency_agency_commences_self-cleansing_.csp

33. Excerpt of remarks by Dauda Garuba at the Public Presentation of a Book on Performance Assessment of the Nigeria Extractive Industries Transparency Initiative (NEITI) organised by the Civil Society Legislative Advocacy Centre (CISLAC) at Bolton White Hotel Abuja, 30 September 2010.

3. DYNAMICS OF INTERACTION AND SYNERGIES

Civil Society and the NEITI Process

The formation of NEITI has brought about certain positive changes to oil and gas revenue governance in Nigeria. Its success in this regard can be measured by the extent to which civil society embraces it. It is expected that NEITI will provide a popular basis for transparency and accountability in governance and reduce corruption to the barest minimum.

Many expect that civil society will mobilise people to put pressure on government to ensure compliance with the open disclosure principles laid down by the NEITI rules. As Ikubaje (2006) suggests, civil society performs this function by dissemination of information about the principles and activities of NEITI to the general public, thereby helping to enhance its legitimacy among the populace. It also involves itself in capacity-building activities to acquire the necessary technical knowledge about the operation of the extractive industry, the nature and type of contractual relations between extractive industries and government, as well as the patterns and types of revenue flows from extractive industry producers to government. CSOs have taken it upon themselves to pursue their monitoring activities, including paying advocacy visits to extractive industries and government institutions concerned with the extractive business (particularly oil and gas). Since 2004, they have also actively participated in a series of NEITI roadshows and media campaigns aimed at sensitising Nigerians to the activities of the organisation and disseminating its audit reports.[34]

Civil society is also expected to be active in monitoring and evaluating NEITI activities to determine whether they meet the standards set at the level of global EITI. This watching of the watchdog is meant to keep it on the alert in performing its functions. By virtue of its representation in the NSWG, civil society is expected to contribute to NEITI policy formulation and programme design. It has done this through active participation in NEITI's Civil Society Working Group/Steering Committee, which came into existence through a memorandum of understanding signed in February 2006.[35] Civil society also has a liaison officer in the NEITI secretariat, whose primary role is monitoring the activities of NEITI and reporting back to the civil society constituency. Finally, civil society has the critical role of disseminating the results of the NEITI audit reports in its campaign for transparency and accountability in the extrac-

---

34. The first round of roadshows across the six geopolitical zones of Nigeria was organised in 2006, while another series ran between 2009 and 2010. What was common to both was the fact that they disseminated and popularised the NEITI audit reports of 1999-2004 and 2005 respectively.

35. The MoU was a compromise between NEITI and civil society over the latter's challenge of the Obasanjo administration's decision to foist a civil society representative on the NEITI board without due consultation.

tive industry. For example, ZCC, a coalition of NGOs, has used the first NEITI audit to engage the NEITI secretariat. CISLAC has also done very well in this regard, by taking key messages to the National Assembly, civil society and the media. Indeed, it remains the only CSO that has consistently engaged on these issues, since many other efforts have either been through coalition groups or one-off events by individual organisations.

There is no gainsaying that civil society has made real progress in advancing the cause of NEITI since 2004. The multi-stakeholder approach is seen as the key to advancing the civil society anti-corruption drive. Through its representation on the multi-stakeholder working group and the appointment of a civil society liaison officer[36] within the secretariat, the NEITI process allows civil society "to be of the secretariat of NEITI but not in the secretariat". This tends to ameliorate what has been seen as one of the major weaknesses of the NEITI Act 2007. The contradiction between NEITI as a government agency and NEITI as a watchdog organisation is one often pointed out as a potential weakness. A government agency watching over another government agency is problematic in a context where criticism of one organisation by another within the same government could be seen as sabotage or betrayal. However, civil society can provide an independent source of domestic and international support and legitimacy for NEITI.

NEITI has been credited with changing attitudes and mentalities in the oil and gas industry in Nigeria. Oil industry operators have become more alert to their responsibilities on issues of accurate disclosures on payments into the federal government coffers, just as various government agencies through which receipts flow or in which they are kept, have also become sensitive to their accountability.

NEITI has also drawn public attention to the scourge of crude oil theft. This is in spite of the fact that the clandestine nature of oil theft (or the illegal oil bunkering business) makes it difficult to come up with reliable statistics. Figure 6 by Stephen James provides an estimate of the volume of crude oil stolen between January 2000 and September 2008.

NEITI has been able to mobilise international pressure in this regard and compel President Umaru Yar'Adua to call on the international community to consider banning "blood oil" the same way it did "blood diamonds":

"I appeal to you and through you to all other G8 leaders to support my new proposal which I will also discuss with UN Secretary General at my meeting with him, that stolen crude should be treated like stolen diamonds because they

---

36. Uche Igwe, a civil society activist, formerly of CLEAN, has occupied this position since its inception.

FIGURE 6. ESTIMATED VALUE OF STOLEN OIL AND SHUT-IN OIL PRODUCTION, IN NIGERIA, JANUARY 2000–SEPTEMBER 2008

Year	Average price of Bonny Light per barrel (in US$)	Volume of oil stolen per day (in barrels	Value of oil stolen per annum (in US$)	Volume of oil shut-in per day (in barrels)	Value of oil shut-in per annum (in US$)	Total value of oil stolen or shut-in per annum (in US$)
2000	28.49	140,000	1.5 billion	250,000	2.6 billion	4.1 billion
2001	24.50	724,171	6.5 billion	200,000	1.8 billion	8.3 billion
2002	25.15	699,763	6.5 billion	370,000	3.4 billion	9.9 billion
2003	28.76	300,000	3.2 billion	350,000	3.7 billion	6.9 billion
2004	38.27	300,000	4.2 billion	230,000	3.2 billion	6.4 billion
2005	55.67	250,000	5.1 billion	180,000	3.7 billion	8.8 billion
2006	66.84	100,000	2.4 billion	600,000	14.6 billion	17.0 billion
2007	75.14	100,000	2.7 billion	600,000	16.5 billion	19.2 billion
2008	115.81	150,000	6.3 billion	650,000	27.5 billion	33.8 billion

Source: Based on figures 8, 9 and 12 in Stephen Davis, *The Potential for Peace and Reconciliation in the Niger Delta*, Coventry Cathedral, February 2009, accessed at www.legaloil.com

both generate blood money. Like what is now known as "blood diamond", stolen crude also aids corruption and violence and can provoke war."[37]

In addition, there is growing awareness of the phenomenon of oil spillage and the necessity of making commensurate reparations to those affected. Previously, oil spillage attracted only token compensation. However, with the coming of NEITI, there has been remarkable improvement in the compensation of victims, with civil society using the opportunity provided by the PIB to press for a 30 per cent share of royalties paid by multinationals to be paid to oil-bearing communities. Extractive industry CSOs also campaign against other unwholesome behaviour by oil and oil services companies, such as the Halliburton bribery scandal that made headlines in Nigeria. Halliburton had long been in Nigeria, but was virtually unknown to the public until a few years ago, when NEITI and its CSO associates publicised its shady dealings, specifically bribing Nigerian government officials to the tune of $180 million to secure a contract to construct a multibillion dollar liquefied natural gas plant in the Niger Delta (between 1995 to 2004).[38]

Challenges

In spite of these achievements, CSOs face a number of challenges in playing their role in the NEITI process. Some of these challenges are owing to gross

---

37. Abdulfattah Olajide "Niger Delta – Bunkering Cartel Behind Militants – Yar'Adua", *Daily Trust* (Abuja), Tuesday, 8 July 2008.

38. For details see Dauda Garuba, "Nigeria: Halliburton, Bribes and the Deceit of 'Zero-Tolerance' for Corruption," available at: http://www.revenuewatch.org/news/news-article/nigeria/nigeria-halliburton-bribes-and-deceit-zero-tolerance-corruption

capacity asymmetries between CSOs and the other institutions in the NEITI process. Other challenges relate to the enabling legal framework of the NEITI process and the character of civil society in Nigeria. CSOs also suffer from capacity deficits, and are weak and fragmented, leading to internal squabbles, personality clashes and poor levels of institutionalisation.

There is evidence of a weak understanding of the NEITI Act even within CSOs. This limits their capacity to participate meaningfully in debates relating to extractive industry/revenue governance issues. The extractive industry, especially its oil and gas sector, is highly technical and requires specialised knowledge and capacity that is lacking in some CSOs. To redress this challenge, between 2007 and 2008 NEITI facilitated the training of 120 civil society representatives by the West Africa Non-governmental Organisation Network (WANGONET) in batches using multi-agency donor funding administered by the World Bank.[39]

CSOs have pointed out that the issue of capacity also affects government institutions. The Nigeria Customs Service, as the ZCC has pointed out, often complains that it lacks the capacity to assess the quantity of oil produced. This kind of knowledge gap makes the assessment of tax, royalties, gas flare penalties, value added tax payable by the oil industry exceptionally difficult.

The independence and quality of civil society's participation in the EITI process was initially hampered by government interference in the selection of civil society representatives on the NSWG. While NEITI's decision to revert to signing the MoU with civil society partly redressed this issue, there is still a strong feeling that the EITI process in Nigeria will run better if civil society is able to elect its representatives to the NSWG, given civil society's potential for feedback and galvanised participation.

There is also the tendency of the NEITI secretariat to view only those CSOs based in Abuja or Lagos as the most relevant to the extractive industry. In fact, those CSOs that are critical and have some real understanding of the issues at the grassroots are located in the oil communities themselves, or conduct their field operations in rural areas. The implication is that NEITI may overlook many organisations that are truly engaged with the people directly affected by oil spills, environmental pollution and degradation. This explains why ERA has, for example, criticised the NEITI Act for being silent on environmental issues, arguing that if it had taken organisations working in the oil-producing communities into consideration at its formation, it would not have missed out on putting the environment squarely in its core mandate (Bassey 2010).

ERA therefore insists that NEITI initiate a process of environmental audit in its templates, and include communities directly involved in the process. This

---

39. Dauda Garuba, one of the authors, participated in the January 2008 iteration of the exercise.

argument is anchored on the principle of EITI itself, which sets only minimum standards and allows implementing countries a free hand (on the basis of the diversity of experiences and the absence of a general template for conducting an EITI audit) to accommodate local challenges and opportunities. This approach, apart from demonstrating the uniqueness and strength of the initiative, has also become an albatross, leading to raging concerns among stakeholders about the absence of specificity in the EITI criteria (such as those that could apply to the form and content of EITI reports). If anything, the approach has affected consistency and stakeholder expectations of the credibility of EITI reports (Garuba 2010).

There is a feeling among some civil society activists that the presidency and National Assembly are the most corrupt levels of government in Nigeria and cannot be relied upon to give Nigerians a strong NEITI Act. One demand echoed by some of the activists interviewed is the need to take NEITI directly to the Nigerian people – students, traders and market women, artisans, workers, communities and farmers – as a social movement. As it is now, NEITI works with city-based NGOs. Politicians are no longer afraid of NGOs, especially those in the cities, but they always bow to the "direct form of people's power as manifested in street engagements".[40] To these activists, it is only at this level that NEITI can act as a catalyst for real progress.

Some CSOs have also made the point that NEITI appears more interested in using civil society to achieve legitimacy in the eyes of international donors, rather than to effect real transparency and accountability in the management of Nigeria's oil revenues. Such validation for NEITI as a government agency only allows it to enhance the external image of the Nigerian government, even when the much expected change remains illusory. The focus on validation, some NGOs maintain, has led to loss of momentum in the core mandate of the initiative.[41] The CSOs are also critical of what they consider the politicisation of NEITI. This has created a situation where CSOs have been more passionate about the mandate of NEITI than those in its bureaucracy. According to one civil society activist, "Until NEITI's activities cause discomfort for someone, it will not go places."

It is precisely for this reason that other civil society activists have asked the following questions: What is civil society doing on the demand side of governance to compel NEITI to live up to expectations? What did CSOs do about activities such as arbitrary oil bloc allocations in the dying days of Obasanjo's government? What is civil society doing about its own internal fractures? The secrecy and haste with which the Obasanjo administration allocated a large num-

---

40. Interview with Celestine Akpo-Bari at Bolton White Hotel, Garki, Abuja, 29 July 2009.
41. Interview with Anthony George-Hill at Bolton White Hotel, Garki, Abuja, 29 July 2009.

ber of oil blocs in its last days violated all known rules of transparency and due process. Yet the CSOs that should have vigorously protested barely whimpered.

As noted earlier, civil society is not entirely free of internal incoherence and anti-democratic tendencies. Internal disagreement in the NGO movement prompted the formation of the Coalition for Accountability and Transparency in Extractive Industry, Forestry and Fisheries in Nigeria (CATEIFFN), which currently runs parallel to PWYP Nigeria, but with a seemingly wider focus. Although the idea to create it had been nursed for a very long time (dating as back as far as 2007, during the climax of the internal wrangling in PWYP-Nigeria), CATEIFFN emerged as an alternative platform for many dissatisfied elements within PWYP-Nigeria after the election that saw the exit of David Ugolor, pioneer coordinator, in December 2008. These elements celebrated his departure, but openly expressed dissatisfaction with his "anointed" replacement.[42]

CSOs seem to have great difficulty in creating and managing coalitions. Internal struggles within civil society coalitions often revolve around personalities, with the result is that civil society is incapable of presenting a united front on issues. This factionalisation is often deliberately encouraged by government to weaken its main critic, given the general absence of strong opposition parties. Creating and exploiting divisions among CSOs has been one strategy of the NEITI secretariat in the past to keep key civil society activists in check.

Bringing in a core constituency: Oil and gas workers and the NEITI process

From the perspective of oil sector workers, NEITI should be seen in the same light as EFCC. However, from the point of view of those who work in the system and are close observers of it, NEITI has not really lived up to its mission as an anti-corruption body. Currently, its main achievement is the consciousness, at least, in the minds of elites, that it exists, a consciousness not shared by ordinary Nigerians living outside cities. However, there is a feeling no political will exists to sustain it, and therefore the earlier momentum generated has fizzled.

Workers are especially looking to NEITI to take on many of the challenges within the extractive industry. For example, there is a belief that a number of bogus memoranda of understanding are circulating within the oil and gas industry and that these form the basis of serious revenue leakages. NEITI is expected to look systematically at all these as the enabling act clearly empowers it to do, but it does not seem to have begun this process.

---

42. The major case against Faith Nwadishi was that she was once a member of staff of David Ugolor's African Network for Environment and Economic Justice (ANEEJ), and that she was never in the forefront of the activities of PWYP-Nigeria. It took the skills, brilliance and maturity of members of the Mohammed Salisu committee that conducted the exercise to calm frayed nerves and secure a middle course that produced the Ms. Nwadishi-led executive for PWYP-Nigeria. Indeed, these were the most trying times for the campaign, as it almost broke-up.

But there are other problems which workers confront on a daily basis, and whose resolution by NEITI would enhance transparency and accountability in the industry. For years, oil companies have based their production figures on unverified volume estimates. With the increasing demand for metering, there is resistance from oil companies. Oil workers note that export terminals still use the dipstick, instead of metering. This is a crude procedure, very outdated and susceptible to manipulation, including illegal bunkering activities. At best, it cannot give a precise measure of the quantity of oil actually exported and is a procedure the same oil companies do not employ in their home countries. The fact that it can be manipulated, oil workers insist, makes the managements of oil-producing companies reluctant to adopt the technologically sophisticated and more accurate metering system. Even if oil companies are ready to concede on metering, they take the view that it should be at export terminals, not flow stations or wellheads. This would mean that the loss of crude, which often occurs before the oil reaches the terminals, will continue. While metering should be a first-line solution, the government's proposal was to appoint an independent monitoring agency for the export terminals. However, the opposition of oil industry workers to an independent monitor, on the grounds that jobs would be lost, scuttled the initiative.

Furthermore, regulators within the industry that ought to ensure standards of transparency are themselves not transparent. This is why oil industry workers have called for an autonomous Petroleum Inspectorate Division in place of the current DPR. The main grouse here is that the DPR is appointed and funded by the government. Beyond this, it is informally funded by the oil majors. Even where there is a suspicion (as there often is) that the latter may have abused production quotas, DPR is usually constrained from acting against the alleged offender. And even where it summons the will to act, the audit it must carry out has to rely on the facilities provided by the oil companies. Indeed, it has also relied on Shell Petroleum Development Company (SPDC)'s helicopter to monitor oil spills. Not even the National Oil Spill Detection and Response Agency (NOSDRA), specifically established to work in this area, has fared any better.

In addition to advocating for an autonomous inspectorate, the workers have put in place their own whistle-blowing policy under which they carry out independent investigations. However, they lack the power to impose sanctions: all they can do is report to the DPR or the federal ministry of labour. This whistle-blowing policy needs to be integrated into the NEITI process.

There are many other areas in the oil and gas sector where the voice and presence of NEITI is critically required. There is the fact of access to information within the country. For example, an oil worker made the point that there is nowhere in Nigeria one can get comprehensive information about the operations of Shell or Chevron in Nigeria. Yet these are the biggest oil and gas producers in the country. To get information about their Nigerian operations, one has to go to their headquarters in Netherlands and the United States. The excuse for this is that it is difficult for them to store and manage such information in Nigeria because of infrastructural limitations, principally the absence of a regular electricity supply. However, it is difficult to see how this excuse can hold, given that many of these organisations ensure their own regular electricity supply for their operations. There is an increased clamour for access to information on foreign oil operations in Nigeria to be made available in the country.

Second, there are allegations that products refined in the country are taken out to the high seas and brought back as imported products. Workers have taken this issue up at the level of the EFCC, the government committee on finance, and the state governors. These are examples of efforts by oil workers to ensure transparency using agencies other than NEITI deemed more likely to achieve results.

It is apparent that the interface between CSOs, oil workers and NEITI is inadequate, although all are represented on the NSWG. Labour claims that NEITI regularly invites them to its meetings, and the workers' representative reports regularly on these meetings. However, the feeling among oil and gas workers is that the exercise is simply "to fulfil all righteousness".[43]

Within labour, the view is that reliance on CSOs to achieve transparency in the oil industry will not work, because such organisations lack the technical know-how to effectively monitor the industry. Second, they view many CSOs as being led by opportunists.[44] In the context of political will on the part of NEITI, it is not difficult to see that oil workers are confronted with an uphill task, despite their willingness to engage with NEITI to promote transparency and accountability in the Nigerian oil industry. The result is that the required synergies between labour represented by oil workers and civil society to monitor and report on the industry have remained largely absent.

Getting NEITI across to the public: The Nigerian press

In the pre-NEITI period, the amount of revenue generated from the oil and gas industry was known only to the operators. The public knew little of what was going on in the industry. In fact, the public was not conscious there was a problem with revenues. However, since NEITI was launched, this situation has largely changed. Great public interest in extractive revenues has been generated and the press has been ably mediating the public discussion.

---

43. Interview with an oil worker.
44. While there may be some truth in this claim, it is not always the case that civil society organisations are "fake." The attempt by oil workers to carpet civil society has also been viewed as one of self-glorification and contempt for CSOs' pivotal role in contemporary society. In fact, the influence of civil society in modern times – including the growing insistence by multilateral institutions that they be represented in the affairs of the world – is such that they are recognised as the "fifth estate of the realm".

The success in publicising NEITI stems from early efforts by NSWG to put a communication strategy in place. This is operated by a media team that complements the work of four other teams within the NEITI structure – technical, legislative, focal and civil society teams. Prior to the enactment of the NEITI Act 2007, the media team comprising Olusegun Adeniyi (of *ThisDay* newspaper, who later became special assistant to late President Umaru Yar'Adua) and Orji Ogbonnaya Orji (formerly of Radio Nigeria) led the process that produced a two-track communication strategy based on a four-phase *Maturity model*: (1) Awareness, (2) Education, (3) Insight and Analysis, and (4) Reform and Remediation.[45] The central objective of the strategy was to facilitate increased comprehensiveness, reliability and integrity through quality information; create access, understanding and analysis through analytical quality; and create a platform for "decision-makers and the public to effectively and appropriately remediate problems identified as part of NEITI audits" (Goldwyn 2005:6). Figure 7 below gives a diagrammatic picture of the NEITI communication strategy spanning three phases in its early days of operation.

FIGURE 7. NEITI COMMUNICATIONS MATURITY MODEL

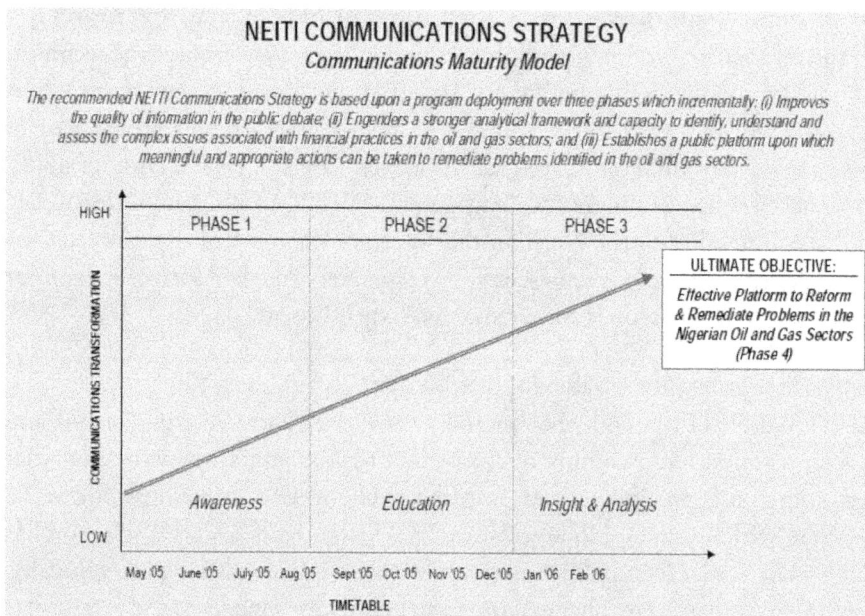

NEITI COMMUNICATIONS STRATEGY
Communications Maturity Model

The recommended NEITI Communications Strategy is based upon a program deployment over three phases which incrementally: (i) Improves the quality of information in the public debate; (ii) Engenders a stronger analytical framework and capacity to identify, understand and assess the complex issues associated with financial practices in the oil and gas sectors; and (iii) Establishes a public platform upon which meaningful and appropriate actions can be taken to remediate problems identified in the oil and gas sectors.

Source: Goldwyn International Strategies (2005), "Nigeria Extractive Industries Transparency Initiatives (NEITI) Communication Strategy." Available at: http://www.neiti.org.ng/files-pdf/NEITI%20Communications%20Strategy.pdf

---

45. See Goldwyn International Strategies, "Nigeria Extractive Industries Transparency Initiatives (NEITI) Communication Strategy", 2005. Available at: http://www.neiti.org.ng/files-pdf/NEITI%20Communications%20Strategy.pdf

Figures 8 and 9 below also diagrammatically represent the two-track implantation strategy (Direct Implementation and Opinion Leaders Implementation) of the NEITI communications maturity model:

The general public can no longer claim to be ignorant about the need for transparency regarding oil revenues and public expenditures, even though the

FIGURE 8. NEITI COMMUNICATIONS MATURITY MODEL: DIRECT IMPLEMENTATION STRATEGY

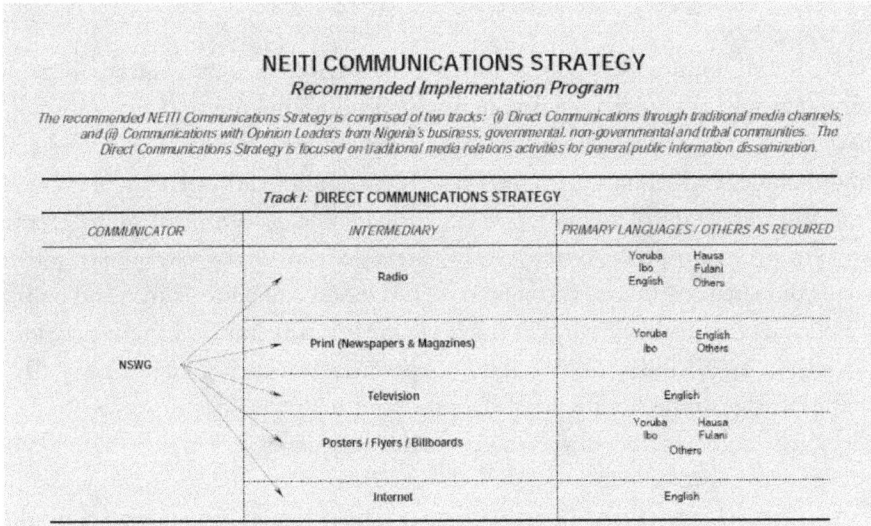

Source: Goldwyn International Strategies (2005), "Nigeria Extractive Industries Transparency Initiatives (NEITI) Communication Strategy." Available at: http://www.neiti.org.ng/files-pdf/NEITI%20Communications%20Strategy.pdf

FIGURE 9. NEITI COMMUNICATIONS STRATEGY: OPINION LEADERS IMPLEMENTATION STRATEGY

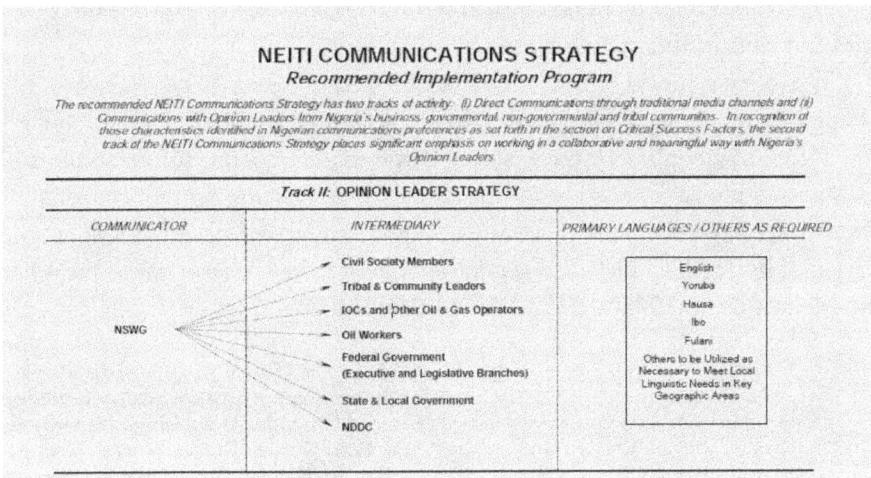

Source: Goldwyn International Strategies (2005), "Nigeria Extractive Industries Transparency Initiatives (NEITI) Communication Strategy." Available at: http://www.neiti.org.ng/files-pdf/NEITI%20Communications%20Strategy.pdf

existing debates remain rather shallow. However, the fact that important questions are now being asked has reduced the information gap, and constitutes an important step forward.

Nonetheless, there are tremendous challenges to meeting the objectives of NEITI. The first is the mentality of those in government, who tend to want to control everything. If government is truly interested in transparency and accountability in the extractive sector, it should allow NEITI to run autonomously, ask pertinent questions and publish its findings, no matter how unpalatable they might be.

This will require three key actions on the part of the government. First, it should ensure adequate (and autonomous) funding for the organisation. Second, the NEITI secretariat must be manned by knowledgeable, committed and highly skilled individuals, with integrity. These qualities are essential for success. It is also important to avoid the general tendency of governments in Nigeria to appoint people to public office largely on the basis of spoils and quotas, rather than competence and commitment to high standards of performance and public service. The grapevine has it that the delay in recruitment into various positions advertised by NEITI since May 2009 is premised on such a flawed basis. This is irrespective of the sensitivity of the position under consideration. The NEITI secretariat must not be allowed to succumb to this logic. Finally, the NSWG must be empowered to do its job. This will entail developing its capacity in extractive industry-related issues. Its role as a board must be insulated from the internal squabbles or external interference usually connected with personality differences, access to resources or the exigencies of the ruling party or corporate actors. The press and CSOs should monitor and report on the NEITI appointment process. The media should also help promote public awareness to pressure government to abide by the statutes in setting up public and governmental agencies and constituting their boards.

The recent removal of Mallam Haruna Sa'eed as executive secretary of NEITI and his replacement by Zainab Shamsuna Ahmed without consultation with the NSWG as required by the enabling law, is a case in point. Information from private and media sources is that Mrs Ahmed is not getting the best cooperation from other NSWG members because her appointment was done in breach of the law establishing NEITI.[46] This is not the first time the Nigerian government has acted like this. In 2008, the Yar'Adua government appointed governing councils

---

46. See Habeeb I. Pindiga, "Jonathan appointed NEITI head illegally," *Daily Trust* (Lagos), Thursday, 5 August 2010. Available at: http://www.dailytrust.com/dailytrust/index. php?option=com_contentandview=articleandid=665:jonathan-appointed-neiti-head-illegallyandcatid=2:lead-storiesandItemid=8; Stan Rerri "Petition: Fraud at NEITI: An Attempt to Silence the Whistle Blower", being a petition to the President of the Federal Republic of Nigeria through the Secretary of the Government of the Federation dated 10 August 2010.

for federal universities without paying attention to the relevant statutes. Only vigorous intervention by the Academic Staff Union of Universities resulted in the reversal of these appointments and adherence to correct procedure.

Getting information to the public about NEITI and the revenue profile of the oil industry has been very slow. This is in part traceable to the audit process, the main basis of getting accurate information on oil and gas revenue, which has been very protracted. The latest audit (2005) is six years out of date. The delay in making the information on oil and gas-derived revenues available through the audit process means that observed deficiencies in revenue-processing procedures in a particular year are not remedied in succeeding years. In other words, errors are repeated simply because the process of unearthing them and proposing remediation is so slow. Nigeria is still very much behind in terms of timeliness and regularity of its EITI audit reports. This problem might be due to the fact that that the audit process is new and complex and developing a reporting template tends to take a particularly long time. However, since two audits have now been done, this need not remain a source of delay.

A number of signals from the oil industry tend to confuse or misinform the public as a result of the lack of up-to-date audit information. For example, during one of those petroleum shortages for which Nigeria has become notorious, a high ranking NNPC official declared there was a shortage of crude oil for domestic use. The very next day, another NNPC spokesman denied there was any crude shortage for domestic refining. Timely audit and an NEITI that is on its toes in terms of field monitoring and information dissemination would have prevented this confusion.

It is also relevant to note that the media face capacity challenges in reporting highly technical oil-industry financial matters. According to a seasoned journalist in this field, 90 per cent of the time journalists rely on information given by the public relations departments of oil firms, the NNPC or DPR. Except for the NEITI-World Bank-sponsored training for civil society and the media in Lagos in 2007-08 and the CISLAC training organised for energy correspondents in Lagos in 2009, little attention has been given to training the journalists who report on the work of NEITI. Investigative reporting that can unearth critical information is hardly undertaken. Thus, while the Nigerian media is very conversant with NEITI, it is not well equipped to engage the NEITI process in a critical and productive manner.

A number of reasons account for this. The first is the risky nature – physical and political – of the terrain. On the physical side, it may be relatively easy to access the savvy public relations departments of oil companies, the NNPC and government ministries, but it is difficult to access oil production and storage fields, which are located onshore or offshore in very rough terrain. This for now is an area for which Nigerian journalists, including those specialising in energy reporting, are not equipped by training or available tools to handle. The political terrain in Nigeria also poses challenges to investigative journalism in the oil sector given the high stakes involved. As an energy reporter made clear, the owners of media organisations are usually close to government and big oil and gas producing companies in what may be considered interlocking relationships. This simply means that media owners are often directly part of the processes of oil-based accumulation that largely defines the Nigerian state and political power. The average reporter is therefore usually very circumspect in the manner s/he reports on the oil industry. In fact, parts of the media resort to self-censorship when they consider a news item too sensitive or likely to upset those in government or the oil corporate sector. At the height of the cultist violence in Rivers State in 2004, a *Daily Independent* correspondent [47] told of his frustration at the "killing" of some of the stories he filed by his editors based on interlocking relationships between his bosses in Lagos and people at the helm in Rivers State. Such unwholesome self-censorship to please people in government compromises reporters and prevents the public from gaining access to vital information.

There are two key areas where the energy desk of the average media house will need to be deployed within the NEITI process, namely the revenue side and the production side. On the revenue side, the skills level within media organisations is reasonably adequate for interactions with the NEITI process. But tracking revenues by matching them with actual production is a level of undertaking that requires different skills and expertise. Feeding the public the necessary information on the NEITI mandate will involve reporting on the very technical, upstream activities of the oil and gas industry, and the extractive industry in general. The knowledge base to ask meaningful questions in this regard is weak and is just not available to the average energy reporter. Filling this gap will require technical training in this specialised area. Also, traversing the rough and risky terrain associated with the oil industry for information will require that reporters have good insurance, currently hardly available to practising journalists in the energy sector, exposing them to a range of field and off-field risks and pressures.

Stakeholders and the National Stakeholders Working Group

NSWG is the board of NEITI, responsible "for the formulation of policy, programmes and strategies" of the organisation. The president is responsible for constituting its membership in accordance with the NEITI Act, which defines the "stakeholdership". While the act stipulates the constituencies from which the membership should be drawn, it does not stipulate the moral, ethical and intellectual capabilities that members must bring onboard. Zonal representa-

---

47. *Daily Independent* is owned by an indigene of the Niger Delta, but published in Lagos.

tion, in spite of its merits for dealing with issues of diversity in Nigeria, puts the NSWG at the mercy of the federal character principle. In accordance with Section 14 (3) and (4) of the constitution of the Federal Republic of Nigeria, this requires that:

> (3) The composition of the Government of the Federation or any of its agencies and the conduct of its affairs shall be carried out in such a manner as to reflect the federal character of Nigeria and the need to promote national unity, and also to command national loyalty, thereby ensuring that there shall be no predominance of persons from a few State or from a few ethnic or other sectional groups in that Government or in any of its agencies.
>
> (4) The composition of the Government of a State, a local government council, or any of the agencies of such Government or council, and the conduct of the affairs of the Government or council or such agencies shall be carried out in such manner as to recognise the diversity of the people within its area of authority and the need to promote a sense of belonging and loyalty among all the people of the Federation.

There are genuine concerns that this provision places the NEITI board and its operations at the mercy of political considerations. Indeed, there is no explicit provision for representatives of the various stakeholders in NSWG to be nominated by their parent constituencies. This makes it possible for the president to appoint supposed representatives, who may not command the respect of their supposed constituencies. This is one weakness that various CSOs see in the manner the NSWG is constituted. Civil society is represented on the board, but CSOs may not be if they do not nominate or elect the person so appointed. In fact, most civil society activists interviewed for this study insist that the civil society representatives on NSWG were not only handpicked by the government, but were also not really part of the extractive industry advocacy movement within civil society. Figure 10 below shows attendance of NSWG meetings by its 15 members since June 2008.[48] The diagram includes representation by proxies, largely restricted to FIRS, Office of the Accountant-General of the Federation (OAGF), and NNPC.

---

48. Members of the NSWG as reconstituted under the NEITI Act by late President Yar'Adua are Professor Assisi Asobie (Chairman), Basil Omiyi (Vice Chairman, Shell), Comrade Shehu Sani, Peter Esele (President, PENGASSAN), Engr. Abubakar Lawal Yar'Adua (GMD, NNPC), Alhaji Jafaru Aliyu Paki, Dr M.I. Yahaya, Alhaji Aliko Mohammed, Leke Alder, T.K. Ogoriba (President, Ijaw World Congress), Mazi Sam Ohuabunwa (President, NESG), Alhaji Ibrahim Dankwambo (Accountant-General of the Federation), Ms. Ifueko Omoigui (Chairperson, Federal Inland Revenue Service), Mallam Mahmud Jega (Editor, *Daily Trust*) and Mallam Haruna Yunusa Sa'eed (Executive Secretary).

FIGURE 10. ATTENDANCE AT NSWG MEETINGS, JUNE 2008–FEBRUARY 2010

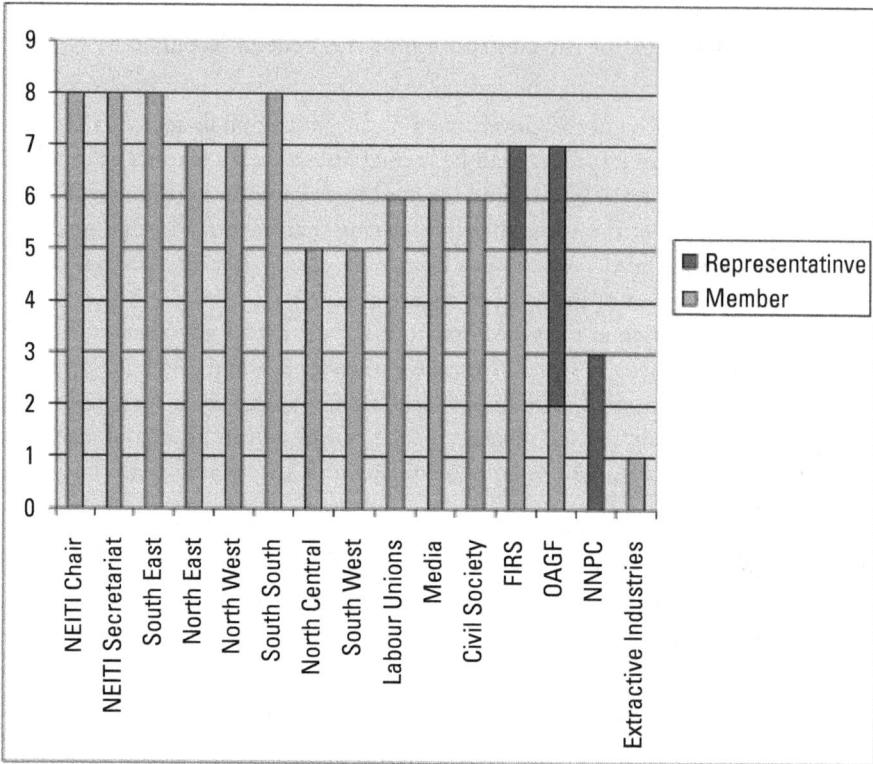

Source: The IDL group, in collaboration with Synergy Global Consulting Ltd (2010), *Validation of the Extractive Industries Transparency Initiative in Nigeria: Final Report,* NEITI Secretariat, Abuja, p.26.

Even if the representations from the extractive industry companies, civil society and labour unions in extractive industries were nominated by their constituencies, they would still constitute only three members of a board of 15 . The others (a chairman, executive secretary, experts in the extractive industry and one member from each of the six geopolitical zones) constitute a solid majority whose appointment to the NSWG may derive essentially from membership of the political party in government. In other words, there are no particular provisions in the act to put NEITI beyond the dominant politics/government of the day. The logic of constituting governing boards in Nigeria is characterised by the handing out of board membership by the president as party patronage to loyalists. As one civil society activist put it, NEITI "is a government agency, and if you are a government agency in Nigeria, you must dance to the tune of the government".[49] This further fuels the feeling that, as currently constituted, NEITI has nothing at stake. The absence of safeguards against possible politi-

---

49. Interview with Emmanuel Ibeh, Stakeholders Democracy Network (SDN).

cally influenced appointments is an obstacle to the anti-corruption thrust of the NEITI process.

This possibility requires specific constituencies to work hard and independently to get their representatives on to the NSWG board, outside the broader political framework defined by party politics and patronage. Already, serious lobbying by oil and gas companies was held responsible for the "watering down" of the NEITI Act 2007. Added to this is the seeming lack of interest in NEITI by the current government. Some civil society activists cynically observe that the only interest government has in NEITI lies in the opportunity for political patronage to loyalists and its potential to impress the outside world that Nigeria is fighting corruption.

NEITI and complementary intergovernmental agencies

The EITI process and the accompanying audit reports have exposed the lack of coordination between the intergovernmental agencies and institutions entrusted with monitoring and managing payments by the extractive industry into government coffers. The 1999–2004 and the 2005 NEITI reports indict the CBN, FIRS and OAGF for the discrepancies in payments by oil companies to the Nigerian government. In the case of the 2005 audit, for instance, the reported figures were higher than what oil companies claimed they paid. Both reports also reveal gaps and weaknesses in NNPC and DPR with regard to their knowledge of the quantity of oil produced by companies, a situation that has fuelled speculation that the country is not getting real value for its products. DPR has neither the tools for precisely measuring the quantity of oil production nor data for estimating, measuring or deducing product losses between production point and terminal. Instead, it relies only on terminal receipts to measure production, which it put at 917.7 million barrels in 2005, or 2.5 million barrels per day .[50] The 2005 NEITI report states that CBN did not record certain Petroleum Profit Tax payments by oil companies amounting to US$241 million, while there are discrepancies in the revenues oil companies paid to the Niger Delta Development Commission (NDDC) and those declared by the commission itself.

The greatest problem associated with the uncoordinated approach of these government agencies that deal with oil and gas matters has to do with the unregulated use of discretionary powers by the leadership of CBN, NNPC, DPR and FIRS. The managing partner of S.S. Afemikhe and Co., Nigeria partner of the Hart Group (UK), recounts how the uncoordinated attitude of these agencies affected transparency and accountability in the oil and gas industry prior to NEITI:

---

50. Bunmi Awolusi, "DPR Doesn't Know Nigeria's Oil Output – NEITI", *Daily Independent* (Lagos) Tuesday, 8 September 2009.

Of course, when you give people discretional power and they know that it ends and closes with them, they do whatever they like. In those days, Federal Inland Revenue Service (FIRS) could end and close the Petroleum Profit Tax (PPT), NNPC could end and close the crude oil sale while Department of Petroleum Resources (DPR) could end and close the royalty. Secondly, the industry is run by divide and rule. If you have a problem with Chevron or Agip, they will run to Nigerian National Petroleum Corporation (NNPC). They know NNPC does not talk to DPR and when they have problem with NNPC, they run to DPR. But now NEITI is now a coordinating body, which puts all these units together and to fight Oil Producers Trade Section (OPTS). I give you an example, this $243 million to $310 million that we talked about was an issue among NNPC, DPR and FIRS. We told all of them to come together led by NEITI, in the FIRS office for discussion and we pointed out to them what we saw that was wrong and they agreed and promised to get the money. Now, we have been speaking with one voice ever since then and that is why this money will come home.[51]

CBN is critical to the NEITI process. As banker to the government, all revenues accruing to the state from whatever source have to be lodged with it. CBN must therefore interact with NNPC, DPR, FIRS and even the oil and gas companies within the framework of NEITI in order to achieve transparency and accountability in revenue tracking and reporting. These interactions are bilateral and multilateral. Bilaterally, the CBN can meet with any of these organisations to compare notes or reconcile accounts. At the multilateral level these organisations meet under the aegis of a committee – Crude Oil Reconciliation Committee (CORC) – which has been set up for that purpose. Receipts are aggregated and accounts reconciled. Indeed, these meetings issue a log table, with well specified objectives, deliverables and timelines for each organisation. Figure 11 below shows in graphic detail what the interaction and reconciliation of intergovernmental agencies and institutions should look like under a well facilitated NEITI process.

NEITI has made substantial contributions to CBN's ability to track resources. Before NEITI, FIRS would lump all revenue receipts into a single account with CBN. With the inception of NEITI, revenue streams from the oil and gas industry have been disaggregated and put into separate accounts to reflect their origin and specific type. Fourteen such accounts for oil-sector revenues are now operated by FIRS with CBN. With this development, it is now very easy to know exactly what accrues from royalties, licensing fees, taxes, etc. With the establishment of NEITI, oil companies now routinely cooperate with CBN to

---

51. Sam Afemikhe Interview in *Nigeria Compass* (Lagos), Monday, 5 April 2010, available at: http://www.compassnewspaper.com/NG/index.php?option=com_contentandview=articlea ndid=45082:oil-industry-is-run-by-divide-and-rule-says-international-auditorandcatid=111 :energyandItemid=712

FIGURE 11. OIL AND GAS REVENUE FLOW CHART

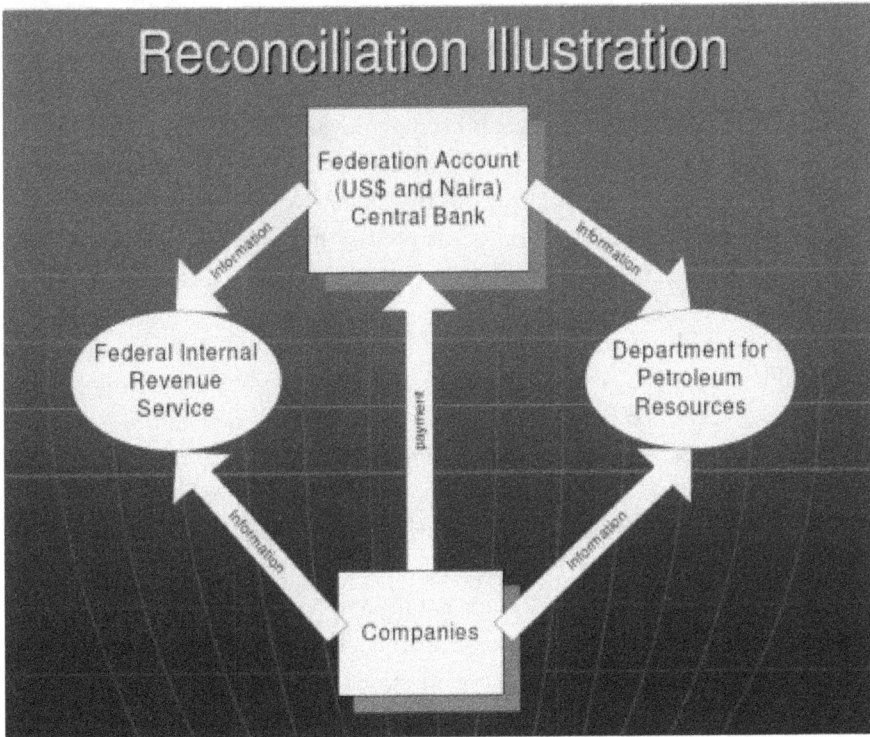

Source: Bright E. Okogu, "Implementing the Extractive Industries Transparency Initiative (EITI): The Nigerian Experience", presentation at a conference on EITI in Ghana – A Success Stoty to be Continued, Accra, 15 January 2007; NEITI, *Nigeria Extractive Industries Transparency Initiative: Audit of the Period 1999–2004 (Popular Version)*, NEITI Secretariat, Abuja, n.d., p.11.

reconcile figures, and the initiative has contributed to advances in oil-revenue transparency beyond intergovernmental organisations. It has also made the oil companies more amenable to scrutiny. FIRS also confirmed this at the high-level roundtable meeting organised by NEITI at the end of first quarter in 2010, where its chair, Mrs Ifueko Omoigui-Okauru, stated that FIRS had undertaken audits and reviews of tax returns of the companies between 1999 and 2004. She added that:

> ... additional assessments have been raised for $605.9m and N2.55bn. To date, payment has been made of $162m and N105m respectively of the assessed amounts ... The objections raised by the oil companies in respect of the out-standing amounts are being resolved and the FIRS are in the process of recovering the outstanding payments. However, some of the issues involved in the objections would require resolution by the courts.[52]

---

52. Obinna Ezeobi, "NEITI, FIRS, Others to End Oil Sector Losses", *The Punch* (Lagos), Friday, 2 April 2010.

To avoid recurring instances of missing records of payment at CBN, Mrs Omoigui-Okauru said FIRS had overhauled the entire process of payment and reconciliation to reflect an electronically-based and fully automated system. The organisation has also worked to address discrepancies in costs and assets provided in the financial statements of oil companies and in their tax returns by designing a template for use by the OPTS with a view to minimising leakages. She added that the capacity of employees of the agency would be improved through technical training and automation of the process and systems.

Challenges

Within the NEITI process, CBN has to take measures to ensure that what it gets tallies with the revenues that others have actually sent in from the oil companies, NNPC or FIRS. NEITI has a reporting template different from the CBN'S template for capturing information. Thus, where one organisation captures information on the basis of a paying company's name while another captures the same information based on what is generated by each oil well, there is bound to be confusion. CBN alleges that this occurred during the first audit and created some discrepancies. There is a need to harmonise the reporting process and agree on a new template for reporting. This should be accompanied by some training to bring the various organisations to a common accounting level with regard to the oil and gas industry. A similar situation should obtain in the interactions between CBN and CSOs in the context of the work of NEITI.

NEITI and the hard road to validation

A major issue that has continued to preoccupy the NEITI secretariat since 2009 has been the validation of Nigeria as an EITI implementing country. Within the lexicon of the EITI implementation process, validation is an overall assessment of a country implementing EITI with a view to ascertaining whether it is compliant or not making meaningful progress. This is EITI's Quality Assurance Mechanism. Put differently, validation is a mechanism that the global EITI board uses to determine a country's candidate or compliant status with a view to protecting the integrity of the initiative.[53]

The purposes of EITI validation are: (1) to enable EITI candidate countries to measure progress in implementation, and (2) to enable EITI compliant countries to undertake an absolute assessment of their compliance with EITI principles and criteria as enshrined in the EITI rules and validation guide. Thus, EITI validation seeks to promote dialogue and learning at the country level and

---

53. EITI, *EITI Rules including Validation Guide*, EITI Secretariat, Oslo, 2010, p.38; The IDL Group and Synergy Global Consulting Ltd, *Validation of the Extractive Industries Transparency Initiative in Nigeria: Final Report*, NEITI Secretariat, Abuja, 2010, p.8.

safeguard the EITI brand by holding implementing countries to the same global standards.

The general approach to EITI validation is set out in the *EITI Rules including the Validation Guide.*[54] The EITI implementing country's workplan, indicator assessment tool and company forms, as well as other evidence of documented information and consultation with stakeholders are used to carry out the validation assignment, through specific approaches and activities that cut across three key stages: (a) preparation, (b) field visits and (c) reporting. The flow chart of the EITI validation process is shown in Figure 12 below.

FIGURE 12. FLOWCHART OF THE VALIDATION PROCESS

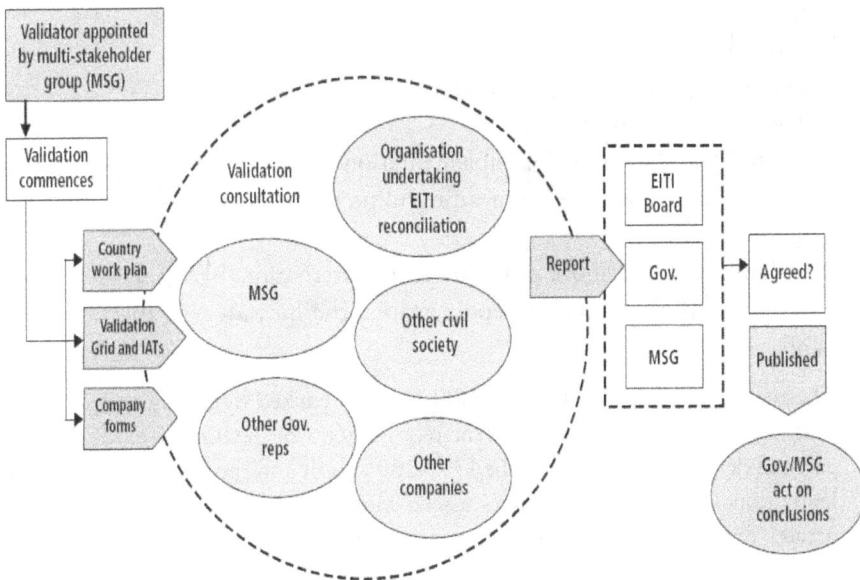

Source: EITI, *EITI Rules including Validation Guide,* EITI Secretariat, Oslo, p.12.

Nigeria opted to go for validation at the Washington DC meeting of EITI, having operated as a candidate country for two years. The NEITI validation contract was awarded to the IDL Group, which collaborated with Synergy Global Consulting Ltd, both based in the UK, to conduct the exercise. The validation exercise covers only oil and gas, given the waiver granted the country on the solid minerals component up to 2012, when it is expected to produce an audit, either separately or together with the oil and gas audit.

Even though the validation report on Nigeria was successfully submitted on 23 February 2010 ahead of the EITI secretariat deadline, the country was (surprisingly) among 15 other candidate countries granted a six month-exten-

---

54. Ibid.

sion to conclude their EITI-compliant status, having demonstrated "exceptional and unforeseeable circumstances outside their control" (Garuba and Ikubaje 2010:153).[55] Our inquiries revealed that Nigeria had applied for extension as it feared it might not meet the March deadline. The real opportunity to audit the EITI implementation in Nigeria came during six month extension granted the country in April 2010.

Nigeria's validation process passed through the necessary procedures defined in the *Validation Guide* that mandates the validator to complete a report comprising:
• A short narrative report on progress against the country workplan;
• A short narrative report on progress against the indicators in the validation grid;
• The completed validation grid;
• A narrative report on company implementation;
• Collated company forms; and
• An overall assessment of the implementation of EITI: is a country a candidate, compliant or is there no meaningful progress?

In accordance with the above procedures, the international EITI secretariat assessed Nigeria's draft validation report against the indicators in the validation grid and concluded:

> The report reflects a trailblazing EITI country backed by strong political support that, at first, vastly exceeded the requirements of the initiative: the unprecedented detail and coverage of the 1999-2004 audit and the first EITI-dedicated legislation. The 2005 audit was also a monumental achievement with even greater revelations and sound recommendations. The process has also clearly played a significant role in achieving some important outcomes, especially improved revenue recovery, platform for increased transparency in government procurement and expenditure, identification of significant improvements to oil and gas management (e.g. well head metering and contract transparency), and extended civil society engagement in policy development. These are important strides.

Although the validators scored Nigeria as having met all indicators, the EITI international board "was not satisfied that the validation report conclusively demonstrated this". It was critical of NEITI in a number of ways, including concerns about the timeliness of NEITI reporting[56] and multi-stakeholder governance process. The board thus recommended the following remedial actions

---

55. Other countries on the list were: Cameroon, the Democratic Republic of Congo, Gabon, Ghana, Kazakhstan, the Kyrgyz Republic, Madagascar, Mali, Mauritania, Niger, Peru, the Republic of Congo, the Republic of Yemen, Sierra Leone and Timor-Leste.
56. Nigeria's latest report was the 2005 audit published in 2009.

for Nigeria to achieve compliance status in a further six months, ending 19 April 2011. As detailed in a *Summary Report* circulated by the international coordinator of PWYP, Radhika Sarin,[57] the actions are:

1. Publication and dissemination of the EITI report covering 2006-2008, to include a clearer definition of materiality for coverage in the report agreed by the NSWG. The NSWG is encouraged to consider a specific figure that defines material payments and revenues, rather than a percentage. This new definition should then be incorporated into the reporting templates for the 2006-08 report;

2. Development and agreement of an NSWG Charter to strengthen oversight of the EITI process;

3. As per the Indicator Assessment Tool for Indicator 13,[58] the government and NSWG should take steps to ensure that all government disclosures to the reconciler are based on audited accounts to international standards and agree a strategy for addressing these issues in accordance with the requirements as specified in Validation IAT [Indicator Assessment Tool] 13;

4. Development, agreement and implementation of an action plan to ensure the comprehensive disclosure of signatory bonuses, and to improve the application of international auditing standards to cover these signature bonuses;

5. Production of a 2005 popular NEITI audit report for a more effective dissemination; and

6. The NSWG develops and agrees a workplan that sets out how the above actions can be accomplished by 19 April 2011. The workplan should also address the steps required by the NSWG to ensure that figures from the Joint Development Zone with Sao Tome e Principe are included in the report as soon as possible.[59]

Notwithstanding the above demands, the EITI board considered it necessary to designate Nigeria as *Close to Compliant*, pending conclusion of the remedial actions by 19 April 2011. Perhaps to arrest Nigeria's disillusionment and ensure prompt completion of the remedial actions, the board agreed, subject to a request by the NEITI secretariat by 15 January 2011, to ensure the review findings would be considered ahead of the EITI global conference billed for March 2011.

Stakeholders in and observers of EITI implementation in Nigeria have been constantly disappointed about the country's seemingly tortuous road to validation, especially as the *Close to Compliant* designation may portend "the default option for countries that have not achieved compliance" – a position the EITI

---

57. Report was circulated on Tuesday, 26 October 2010.
58. ndicator 13 asked: "Has the government ensured that government reports are based on audited accounts to international standards?"
59. Publish What You Pay, "Summary Report: 13th EITI International Board Meeting held in Dar es Salaam", 19-20 October 2010.

board has denied.[60] The situation is underpinned by their knowledge of the country's global leadership role in the process prior to the end of the Obasanjo administration in May 2007. This, if anything, points to the personality factor in public policy actions and governance, especially as it has been observed that the current EITI process in Nigeria lacks the commitment that was accorded it during Obasanjo's administration. Concerns about the Yar'Adua-Jonathan government's commitment is further reflected in the fact that the validators were unable to obtain feedback from many ministries, departments and agencies invited for interviews. The final validation report has it that:

> … the validators explain that they sought meetings with the DPR, the Central Bank, the Accountant General and the Revenue Mobilisation, Allocation and Fiscal Commission without success, even inviting stakeholders to provide comments by public notice. The validators did succeed in meeting with FIRS and NNPC …[61]

Perhaps, that sense of commitment is what the recent appointment of Mrs Zainab Ahmed as NEITI executive secretary is expected to renew. Already, a special task force on validation has been set up at a recent NSWG meeting with the sole objective of ensuring that Nigeria meets all the validation conditions set for it by the EITI board by 15 December 2010. The task force has hit the ground running, with its alternate chair, Mazi Sam Ohuabunwa, disclosing "that responses from relevant government agencies expected to supply data on revenue flow from the oil and gas for the NEITI Audit have been positive". Ohuabunwa identified the agencies as CBN, FIRS, DPR, OAGF and NNPC, among others.

The new executive secretary, Mrs Ahmed, has further assured Nigerians that with the support of civil society, companies and government agencies involved in the implementation of the workplan on each of the six conditions of remediation, NEITI is set to meet the EITI board by an earlier date of 15 January 2011, adding that: "Our target is even to beat the April 2011 date given to NEITI by the world body". Nigerians look forward to this new promise of commitment, and the rest of the story will only unfold with time.

---

60. See EITI, "Validation: Nigeria", Board Paper13-3-D, EITI Validation Committee, EITI Secretariat, Oslo, 4 October 2010. Close to Compliant is one of two new phrases adopted at the Dar es Salaam board meeting in October 2010 to categorise EITI candidate countries, the other being Meaningful Compliant.

61. This is contained in the EITI secretariat's Assessment of the Final Report column in: The IDL Group and Synergy Global Consulting Ltd (2010), Validation of the Extractive Industries Transparency Initiative in Nigeria: Final Report, NEITI Secretariat, Abuja.

4. NEITI AND EMERGING ISSUES IN THE EXTRACTIVE INDUSTRY

The Petroleum Industry Bill and NEITI

The PIB, which has been under consideration by the National Assembly for some time, has the potential to reinforce NEITI's ability to make oil companies more accountable to local oil-bearing communities. So far, the evidence suggests PIB does not contain provisions to this effect. As part of the amnesty deal with militants of the Niger Delta, the government of President Yar'Adua proposed divesting 10 per cent of its equity share in oil and gas to oil-producing communities in the delta This proposal was widely acclaimed – in spite of the confusion regarding the modalities for sharing the dividends among communities so defined – and was expected to be incorporated into the PIB.[62] This will give local oil-producing communities a direct stakeholder role in NEITI. If this proposal is incorporated into the PIB, oil-producing communities will have a direct 10 per cent revenue stake in oil investments, and will have to work with NEITI to access to accurate information on oil and gas revenues from oil companies and ensure that they get their due entitlements.

PIB may need to involve NEITI's oversight role in the transition from JVCs to IJVCs. The current system of oil production in Nigeria involves the Nigerian government forming JVCs with individual oil and gas producing companies. Expenses, especially exploration and production costs, are shared according to equity stakes. These are paid at the beginning of every year. For the Nigerian government, the annual joint venture cash call runs into huge sums in US dollar terms and is usually a source of real pressure on the annual national budget. On the other hand, oil and gas companies have always used their oil bloc acreages and reserves as collateral to obtain relatively cheap loans to meet their cash call obligations, while the Nigerian government simply sits back without contributing its share of money, except to wait for it to be deducted from whatever it accrues from oil sales and profits from exports. Figure 13 below shows an oil and gas production flowchart.

---

62. The Senior Special Assistant to the President on Petroleum Matters, Dr Emmanuel Egbogah, has been canvassing this idea, which he espoused at an International Conference on *Fuelling the World – Failing the Region? New Challenges of Global Energy Security, Resource Governance and Development in the Gulf of Guinea* organised by Friedrich-Ebert-Stiftung at the Nicon Luxury Hotel, Abuja, Nigeria from 25–26 May 2010. Dr Egbogah's presentation raised many questions and comments from the audience that could not be fully addressed. For most recent works on the 10 per cent proposal, see Aaron Sayne (with contribution from Jim Williams), *Antidote to Violence? Lessons for the Nigerian Federal Government's Ten Percent Community Royalty from the Oil company Experience*, Niger Delta Report, No. 1, Transnational Crisis Project, Washington, DC, February 2010; Aaron Sayne (with contribution from Jim Williams), *Something or Nothing: Granting Niger Deltans a Stake in Oil to Reduce Conflict*, Nigeria Policy Paper, Transnational Crisis Group, Washington, DC, October 2010.

FIGURE 13: OIL AND GAS PRODUCTION FLOWCHART

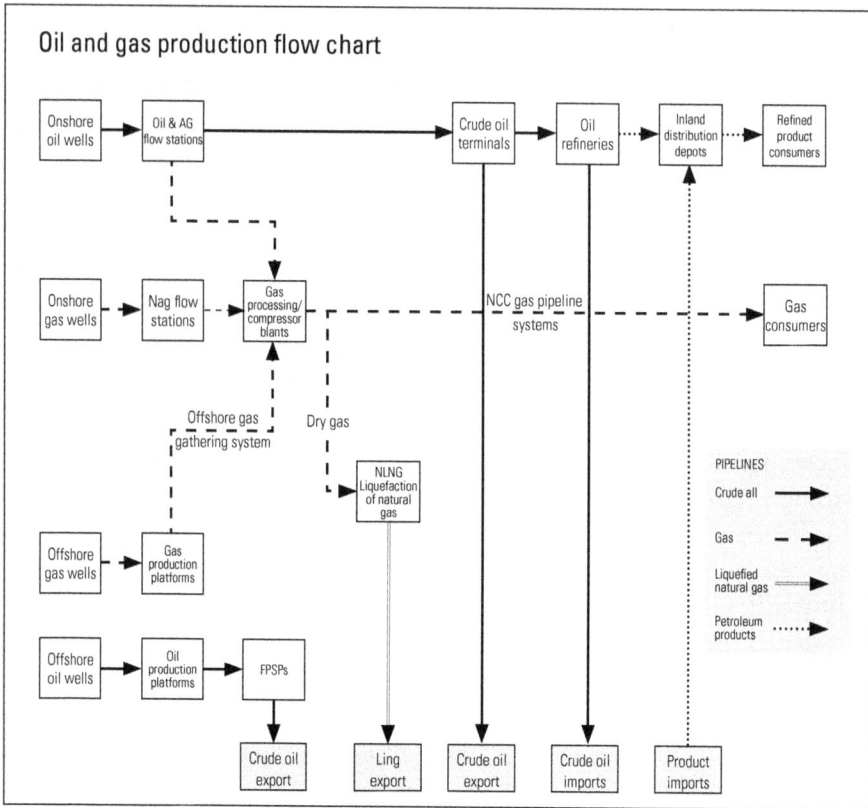

Oil and gas production flow chart

Source: NEITI, Nigeria Extractive Industries Transparency Initiative: Audit of the Period 1999–2004 (Popular Version), NEITI Secretariat, Abuja, n.d., p.18.

PIB seeks to correct the imbalances in the JVC partnership between government and each IOC by converting it into an IJVC. The IJVC is then expected to use its oil blocs and reserves as collateral to meet cash call obligations. This will free the government from the usual pressure on the annual budget. While this proposal has remained controversial and constitutes one of the main points of disagreement between government and oil companies, if it sails through it is likely to pose new challenges for the NEITI process. As with other financial resources involving the government, NEITI may have to capture and monitor them. This will demand that NEITI take on challenges requiring sophisticated technical skills that it may lack in its current state. When enacted, it is expected the PIB will address many of the issues that surround the licensing of oil blocs and memoranda of understanding with local communities, among others, all of which the NEITI process audit has revealed to be far from transparent under the current oil regime.

The delay in passing PIB into law may stem from lobbying activities or pressures on lawmakers by IOCs or the OPTS of the Chamber of Commerce and Industry. It is therefore important that NEITI and its stakeholders follow up on the deliberations on the bill, beyond the memorandum it presented at the public hearing at the National Assembly, to ensure that its oversight functions are reinforced by the law.

NEITI and the solid minerals sector

There has been a strong demand for NEITI to expand its work into the solid minerals sector to remedy the obvious neglect of the sector and the associated monetary losses. This sector represents the oldest mineral exploitation business in Nigeria, dating back to the opening of the Enugu mines in 1915. However, the country's shift to an oil monoculture has relegated the solid minerals sector to the background. Not even promises by successive governments (including the creation of a ministry of solid minerals development by the Abacha regime [1993-98]) to diversify Nigeria's economy have been matched by serious action. The result is continued illegal exploitation of solid minerals across the country.

One of the few major attempts to get the Nigerian government to open up its solid minerals development sector was the World Bank-supported Sustainable Management of Mineral Resources Project, through which the country secured a US$120million loan facility in 2006 to expand and develop the sector. The facility – interest free, with a 35-year term and 10-year moratorium – enabled the federal government to establish a cadastral office, provide loans to artisanal miners, establish a legal regulatory framework (solid minerals development policy and Mining Act 2007) and revalidate the mining cadastre. The result has been the granting of the first batch of 1,002 mineral title licences in May 2007 (Garuba and Ikubaje 2010:151). NEITI has also established a database on mining permits and licensing (Ezekwesili 2006).

While nothing much has been heard about progress since 2007, the granting of another US$80million facility to Nigeria in 2011 on grounds that the country satisfactorily utilised the earlier d $120m loan indicates continued interest in getting Nigeria to attend to its solid minerals sector. Also, the request by the EITI board at a meeting in Washington DC that Nigeria move ahead to audit the solid minerals sector by the end of 2012, whether as a separate report or part of an overall report, could be seen as the beginning of a process that is long overdue. With over 30 variants of solid mineral deposits across the country, Nigeria is expected to derive as much from other natural resources as it gets from oil, if not more, especially with the continued pressure on the federal government to explore alternative sources of income besides crude oil and gas.

5. CONCLUSION

Summary of findings

Nigeria has come a long way in implementing EITI since it signed on to it. Given the country's enthusiasm in embracing the transparency initiative, observers in and outside the country are concerned by the decline in Nigeria's rating, particularly after it suffered two setbacks in the validation process in March and October 2010.

Key policy recommendations

The following policy recommendations are made to address most of the concerns raised by observers with regard to the perceived decline in Nigeria's rating. They are aimed at further empowering the NEITI process as a strategy for institutionalising transparency and accountability in the governance of the extractive resources sector in Nigeria.

Building capacity among stakeholders

NEITI has brought together various stakeholders in the extractive resources industry, but most of them have little understanding of that industry. CSOs, media practitioners, community activists and NEITI workers all need serious retooling on the oil and gas industry.

Declaring a NEITI awareness week

There is good awareness of the NEITI process among CSOs, labour unions and oil workers. But this awareness tends to be restricted to elites in urban centres and leaves out most of the people living in the rural areas, where most of the extraction of natural resources occurs and where people's livelihoods and living environments are directly affected by mining and oil company activities. There is an urgent need to popularise the message of NEITI to students, market women, artisans and the majority of Nigerians living in rural and urban areas. An annual *NEITI week* should be declared, with programmes targeting the public in all parts of the country. This should include the use of radio and social media to sensitise people to the role of NEITI and their responsibility to ensure it works to promote transparency and accountability in the country.

Expanding the transparency component of the NEITI process

There is a need to expand the concept of transparency in the revenue inflow process by making relevant government bodies publish what they receive as revenues while the paying companies simultaneously publish what they have paid into government accounts. Such information should also be communicated to the NEITI secretariat as the payments are made and receipts acknowledged. This is important in making public the reconciliation of accounts with regard to amounts paid, by whom, and the receipts made for such payments by the receiving agency.

Pushing the expenditure side of NEITI

It seems the framers of EITI assume that access to information and transparency will automatically translate into benefits for citizens of resource-rich countries. However, EITI has become a vehicle for promoting revenue transparency while largely ignoring the expenditure side. This perhaps explains why Nicholas Shaxson (2009) asked if the NEITI audit was not "just a glorious audit". With "unreceding" poverty in the land, there are growing fears the initiative could lose its essence and thereby advance the argument that EITI is not a solution to the "resource curse". Thus, to ensure that transparency becomes a means for poverty reduction and the achievement of the good things of life for citizens of resource-rich countries, there is the need to expand the scope of NEITI (as has also been advanced for EITI) to include the expenditure side.

Election of some members of the NEITI board

There is also the problem of the composition of the NEITI board. The president has virtual authority to appoint members of the board. Presidential appointment of members of civil society (NGO, labour and media representatives, as is currently the case) works against true representation of civil society. It is important that civil society representatives on NSWG be directly elected by their own constituencies. This will serve to guarantee their autonomy and encourage consultation with and accountability to their constituencies about their activities on the NEITI secretariat, rather than seeing themselves as responsible to the president who appointed them.

Taking the NEITI process to other levels of government

There is a need to extend the NEITI process to other levels of government. Nigeria operates a federal system in which substantial resources are devolved to states and local government councils. Currently, states and local governments are disconnected from the NEITI process, yet payments such as income taxes by extractive industries to state governments should fall within the ambit of the NEITI audit. The fact that oil is the revenue mainstay of states and local governments and that the principle of derivation grants 13 per cent of oil and gas revenues to producing states should give all of them a stake in ensuring the accuracy of what goes into the federation account from which allocations are made. It is necessary to integrate states and local communities into the EITI process at two levels. The first is to initiate the process at state and local levels. The second is to bring in the states and local communities into the NEITI process as stakeholders. This will require an amendment to the NEITI Act 2007.

Incorporating downstream revenues into the NEITI process

NEITI is mostly concerned with revenue from the upstream sector of the oil industry. Yet there are significant revenues derivable from downstream activities that are not covered in the NEITI Act 2007. And with PIB in the making, it is most likely the midstream sector will also generate activities that will cross the jurisdiction of NEITI. To get the full benefit of the entire value chain, the downstream revenues need to be included in the revenue profile of the oil industry

Establishing NEITI units in extractive companies

Every oil company should be compelled by law to have a NEITI unit to keep track of payments the company makes to all levels of governments (national, state and local councils) as well as communities. This unit should also form the basis of continuous interaction with the broader NEITI process and will also help in the development of technical knowledge of the extractive sector, currently in a deplorable state.

Raising the profile of the demand-side of governance

NEITI may be the supplier of information, but there has to be informed demand from civil society. There is a need to expose CSOs to the technical terms in the production and audit processes so they can make more informed commentaries on these processes. Legislation is also needed to facilitate this type of capacity building to empower civil society to exercise its oversight functions in ensuring transparency and accountability in the extractive industry. To this end, the legislature should be sensitised to the importance of its role in facilitating the capacity of civil society as a way of ensuring the overall success of NEITI. As currently constituted, the two chambers of the National Assembly have functions related to the NEITI process scattered across committees – oil, gas and solid minerals. It is important that these activities be coordinated for greater coherence and effectiveness.

Creating a civil society programming unit within the NEITI secretariat

There is also a related need to create a civil society programming unit in the NEITI secretariat. Currently, civil society is not getting full value from its representation through the civil society liaison officer, Uche Igwe. As up-to-date as he currently is on EITI issues, he is hamstrung in his role because of the weak civil society programming content in the NEITI secretariat. Much is required of him by way of feedback to the civil society constituency with a view to strengthening and ensuring regular NEITI-civil society working group interaction This unit should be responsible for developing a civil society-driven agenda within NEITI and creating the necessary points of contact between CSOs and other stakeholders in the NEITI process. Such a unit needs to be strongly gender-sensitive, an attribute apparently lacking at present throughout NEITI.

Pushing for a new international financial reporting standard for natural resource industries

There is an ongoing push for a new global standard for extractive company reporting. The initiative, if adopted, has the potential to make oil, gas and mining companies publish what they pay to the government of each country in which they operate as well as what they extract, including costs of production, production revenues and the reserves for each country. Advocates of transparency and accountability around the world agree that mandating companies to publish this information would be an enormous breakthrough in the global struggle for transparency and accountability in natural resource extraction, given its potential to create a global standard of disclosure that would provide citizens with critical information required to hold their governments accountable. The movement is organised around the International Financial Reporting Standards (IFRS), which epitomise principles-based standards and interpretations, and the framework adopted by the International Accounting Standards Board (IASB) representing about 100 countries in Africa, Europe and elsewhere.

While CBN and the Securities and Exchange Commission (SEC) favour adoption of the standards from December 2012, experts in the Nigerian Accounting Standards Board say there is no need to rush (Ahmed 2010). Nigeria can be part of the global groundswell of mobilisation to ensure that domestic natural resources are a focus in development financing in resource-rich countries as well as promoting fiscal transparency in tax revenues from extractive companies by supporting the IASB.

BIBLIOGRAPHY

Aderinokun, Kunle (2010) "Crisis Rocks NEITI Over Allegation of Corruption", *ThisDay* (Lagos), Tuesday, 24 August.

Ahmed, Idris (2010) "Is Country Prepared for International Financial Reporting Standards?", *Daily Trust* (Abuja), 29 August.

Ajayi, Kunle (2003) "Bureaucratic Corruption and Anti-Corruption Strategies in Nigeria, 1976–2001", in Godwin Onu (ed.), *Corruption and Sustainable Development: The Third World Perspective,* Abuja and Awka: IPSA Research Committee 4 on Public Bureaucracy in Developing Societies and Nnamdi Azikiwe University, pp 153–72.

Asobie, Assasi (2010) "Re-Petition: Fraud at NEITI: An Attempt to Silence the Whistle Blower" (Ref: NEITI/ADM/079/Vol.2/242), a 14-page letter to the Secretary to the Government of the Federation, Alhaji Mahmud Yayale Ahmed, being a response to Mr Stan Rerri's petition to President Goodluck Jonathan, 19 August.

Asuni, Judith Burdin (2009) "Blood Oil in the Niger Delta", *Special Report 229,* Washington DC: United States Institute for Peace.

Awolusi, Bunmi (2009) "DPR Doesn't Know Nigeria's Oil Output – NEITI", *Daily Independent* (Lagos), Tuesday, 8 September.

Bassey, Nnimmo (2010) "The Environmental Black Hole in NEITI", in Sofiri Joab-Peterside, Ekanem Bassey and Naomi Goyo (eds), *Domestication of Extractive Industries Transparency Initiative in Nigeria: Gaps between Commitment and Implementation – A Civil Society Assessment of the Performance of Nigeria Extractive Industries Transparency Initiative,* Abuja: Civil Society Legislative Advocacy Centre, pp. 93-109.

Biang, J. Tanga (2010) "The Joint Development Zone between Nigeria and Sao Tome and Principe: The Case of Joint Development in the Gulf of Guinea – International Law, State Practice and Prospects for Regional Integration", New York: Report of United Nations-The Nippon Foundation Fellowship Programme 2009/10 submitted to the Division for Ocean Affairs and the Law of the Sea Office of Legal Affairs, United Nations.

Civil Society Legislative Advocacy Centre (2010) "Report on the Public Presentation of Performance Assessment of the Nigeria Extractive Industries Transparency Initiative (NEITI) by the Civil Society Legislative Advocacy Centre (CISLAC) at Bolton White Hotel Abuja, On September 30, 2010".

Civil Society Working on Extractive Revenue Transparency, Accountability and Good Governance in Nigeria (2009) "A Memorandum on the petroleum Industry Bill 2009 submitted to the House of Representatives."

Davis, Stephen (2009) *The Potential for Peace and Reconciliation in the Niger Delta,* Coventry Cathedral, February 2009, accessed at www.legaloil.com

Economic Confidential, (2008) "Accruals in Respect of Signature Bonus from 1999 – 2007" *Economic Confidential,* available at: http://www.economicconfidential.com/febfactdpr.htm.

Esiedesa, Obas (2010) "Petroleum companies yet to remit N345b to FG – NEITI", *Daily Independent* (Lagos), Monday, 6 December.

Extractive Industries Transparency Initiative (2010) *EITI Rules including the Validation Guide*, Oslo: EITI Secretariat.

Extractive Industries Transparency Initiative (2010) "Assessment of the Nigeria Draft Validation Report", Oslo: International EITI Secretariat, 5 March.

Extractive Industries Transparency Initiative (2010) *EITI Fact Sheet*, EITI Secretariat, Oslo, 25 November.

Ezekwesili, Obiageli (2006) "Solid Minerals: A Strategic Sector for National Development", paper presented at the NEITI North Central Road Show at the Transcorp Hilton Hotel, Abuja, 11 April.

Ezeobi, Obinna (2010) "NEITI, FIRS, Others to End Oil Sector Losses", *The Punch* (Lagos), Friday, 2 April.

Federal Government of Nigeria (1999) *Constitution of the Federal Republic to Nigeria*, Abuja: Federal Government Press.

Fred-Adegbulugbe, Chinyere (2010) "Oil workers threaten showdown with FG over PIB", *The Punch* (Lagos) Monday 27 September, p.17. Also available at: http://www.punchng.com/Articl.aspx?theartic=Art201009278373448

Garuba, Dauda (2009) "Nigeria: Halliburton, Bribes and the Deceit of 'Zero-Tolerance' for Corruption", available at: http://www.revenuewatch.org/news/news-article/nigeria/nigeria-halliburton-bribes-and-deceit-zero-tolerance-corruption.

Garuba, Dauda (2010) "Is there a Need for 'EITI Reloaded'? An Assessment of the EITI Process," paper presented at the International Conference on *Fuelling the World – Failing the Region? New Challenges of Global Energy Security, Resource Governance and Development in the Gulf of Guinea* organised by Friedrich-Ebert-Stiftung, from 25-26 May Nicon Luxury Hotel, Abuja.

Garuba, Dauda S. and John G. Ikubaje (2010) "The Nigeria Extractive Industries Transparency Initiative and Publish What You Pay Nigeria", in Mary McNeil and Carmen Malena (eds), *Demanding Good Governance: Lessons from Social Accountability Initiatives in Africa*, Washington DC: World Bank, pp.137–62.

Goldwyn International Strategies (2005) "Nigeria Extractive Industries Transparency Initiatives (NEITI) Communication Strategy", available at: http://www.neiti.org.ng/files-pdf/NEITI%20Communications%20Strategy.pdf

Goldwyn, David L. (ed.) (2008) *Drilling Down: The Civil Society Guide to Extractive Industry Revenues and the EITI*, New York: Revenue Watch Institute.

Hassan, Turaki A. (2010) "PIB: N/Assembly caves in to oil majors, The Jonathan Connection", *Daily Trust* (Abuja), Friday, 8 October.

Humphreys, Macartan, Jeffrey Sachs and Joseph Stiglitz (2007), "Introduction: What is the problem with Natural Resource Wealth", in M. Humphreys, J. Sachs and J. Stiglitz (eds.), *Escaping the Resource Curse,* New York: Columbia University Press.

IDL Group and Synergy Global Consulting Ltd (2010) *Validation of the Extractive Industries Transparency Initiative in Nigeria: Final Report,* Abuja: NEITI Secretariat.

Iyayi, Festus (2000) "Oil Corporations and the Politics of Community Relations in Oil Producing Communities", in Committee for the Defence of Human Rights, *Boiling Point: A CDHR Publication on the Crisis in the Oil Producing Communities in Nigeria*, Lagos: Committee for the Defence of Human Rights.

Kar, Dev, and Devon Cartwright-Smith (2010) *Illicit Financial Flows from Africa: Hidden Resource for Development*, Washington DC: Paper for the Global Financial Integrity Program of the Center for International Policy. Also available at: http:// www.gfip.org/storage/ gfip/documents/reports/gfi_africareport_web.pdf.

Lubeck, Paul M., Michael J. Watts, and Ronnie Lipschutz (2007) "Convergent Interests: U.S. Energy Security and the 'Securing' of Nigerian Democracy", *International Policy Report*, Washington DC: Center for International Policy. Also available at: http:// ciponline.org/NIGERIA_FINAL.pdf.

Müller, Marie (2010) *Revenue transparency to mitigate the Resource Curse in the Niger Delta? Potential and reality of NEITI*, Occasional Paper V, Bonn: Bonn International Centre for Conversion.

Niger Delta Budget Monitoring Group (2009) "A Memorandum on the Petroleum Industry Bill (PIB)" submitted to the Senate and House of Representatives.

Nigeria Extractive Industries Transparency Initiative (2007) *NEITI Act 2007*, Abuja: Federal Republic of Nigeria.

Nigeria Extractive Industries Transparency Initiative (n.d.) *Nigeria Extractive Industries Transparency Initiative: Audit of the Period 1999–2004 (Popular Version)*, Abuja: NEITI Secretariat.

Ogbodo, John-Abba (2009) "Reps ask oil firms to pay N225.45 trillion", *The Guardian* (Lagos), Wednesday, 25 February, available at: http://speakersoffice.gov.ng/news_feb_24_09_1.htm)

Okogu, Bright E. (2007) "Implementing the Extractive Industries Transparency Initiative (EITI): The Nigerian Experience", a presentation at a conference on EITI in Ghana – A Success Story to be Continued, Accra, 15 January.

Okonjo-Iweala, Ngozi and Philip Osafo-Kwaako (2007) *Nigeria's Economic Reforms: Process and Challenges*, Working Paper No. 6, Brookings Global Economic Development, Washington DC: The Brookings Institution.

Olajide, Abdulfattah (2008) "Niger Delta – Bunkering Cartel Behind Militants – Yar'Adua", *Daily Trust* (Abuja), Tuesday, 8 July.

Omeje, Kenneth (2005) "Oil Conflict in Nigeria: Contending Issues and Perspectives of the Local Delta People", *New Political Economy*, Vol. 10, No. 3, September, pp.321–34.

Orogun, Weneso (2009) "NEITI Awards Meter Infrastructure Study", *ThisDay* (Lagos) Thursday, 27 August.

Pindiga, Habeeb I. (2010) "Jonathan appointed NEITI head illegally", *Daily Trust* (Lagos), Thursday, 5 August, available at: http://www.dailytrust.com/dailytrust/ index.php?option=com_contentandview=articleandid=665:jonathan-appointed-neiti-head-illegallyandcatid=2:lead-storiesandItemid=8;

Publish What You Pay (2010) "Summary Report: 13th EITI International Board Meeting held in Dar es Salaam", 19–20 October.

Publish What You Pay and Revenue Watch Institute (2006) *Eye on EITI: Civil Society Perspectives and Recommendation on the Extractive Industries and Transparency Initiative,* London and New York: Publish What You Pay and Revenue Watch Institute.

"Re: Investigation of Payments to Hotels for NEITI 2009 Civil Society Training", Report of the Committee set up by the Nigerian Multi-stakeholder Working Group to investigate payments by NEITI Secretariat to Hotels for the 2009 Civil Society Training, 19/05/10, (unpublished).

Rerri, Stan (2010) "Petition: Fraud at NEITI: An Attempt to Silence the Whistle Blower" (Reference: NENTI/PETITIONS/DS/01), being a petition to the President of the Federal Republic of Nigeria through the Secretary of the Government of the Federation dated 10 August.

Sayne, Aaron (with contribution from Jim Williams) (2010) *Antidote to Violence? Lessons for the Nigerian Federal Government's Ten Percent Community Royalty from the Oil company Experience*, Niger Delta Report No. 1, Washington DC: Transnational Crisis Project.

Sayne, Aaron (with contribution from Jim Williams) (2010) *Something or Nothing: Granting Niger Deltans a Stake in Oil to Reduce Conflict*, Nigeria Policy Paper, Washington DC: Transnational Crisis Group.

Shaxson, Nicholas (2009) *Nigeria's Extractive Industries Transparency Initiative: Just a Glorious Audit?* London: Chatham House.

The Guardian (Lagos), (2008) "65 per cent of Oil Signature Bonus Remains Unaccounted", Wednesday, 30 July, available at: http://www.financialnigeria.com/NEWS/news_item_detail_archive.aspx?item=2746

Udo, Bassey (2008) "FG Demands $231 Million Signature Bonus From Korean Firm for Two Oil Blocks", *Daily Independent* (Lagos), Wednesday, 3 November. Also available at: http://allafrica.com/stories/200811040904.html

Udo, Bassey (2010) "Corruption Allegation dogs Nigeria's Extractive Industry Monitors", *Next* (Lagos), Saturday, 28 August.

Udo, Bassey (2010) "Transparency Agency Commences Self-cleansing" *Next* (Lagos), Sunday, 17 October, available at: http://234next.com/csp/cms/sites/Next/Money/5630854-147/transparency_agency_commences_self-cleansing_.csp

APPENDICES

List of interviewees/Organisations represented at FGDs

A. Interviews

Afemikhe, Sam S., Managing Director, S.S. Afemikhe Consulting (Hart Group), Lagos, 3 August 2009.

Akpatason, Peter O., National President, National Union of Petroleum and Natural Gas Workers (NUPENG), Lagos, 4 August 2009.

Akpo Bari, Celestine, Community Officer, Social Development Integrated Centre, Port Harcourt. (Interviewed in Abuja, 29 July 2009).

Aroyehun, Gbenga, Economic and Financial Crimes Commission (EFCC), Abuja, 28 July 2009.

Ekeanyanwu, Lilian, Head, Technical Unit on Governance and Anti-Corruption (TUGAR), The Presidency, Abuja, 30 July 2009.

Egbule, Peter, West African NGO Network, WANGONET, Lagos, 3 August 2009.

Ereba, Patrick, Program Manager, Centre for Social and Corporate Responsibility, Yenagoa, Bayelsa. (Interviewed in Abuja, 28 July 2009).

Etchere. Eric J., Director, S.S. Afemikhe Consulting (Hart Group), Lagos, 3 August 2009.

George-Hill, Anthony, National Co-ordinator, Niger Delta Budget Monitoring Group (NDEBUMOG), Port Harcourt. (Interviewed in Abuja, 30 July 2009).

Ibeh, Emmanuel , Stakeholders Democracy Network (SDN), Port Harcourt. (Interviewed in Abuja, 29 July 2009).

Igwe, Uche, Civil Society Liaison Officer, NEITI Secretariat, Abuja, 30 July 2009.

Ikubaje, John (PACT), Abuja, 1 August 2009.

Lardner, Tunji, West African NGO Network, WANGONET, Lagos, 3 August 2009.

Mairiga, Umar A., Assistant Director, Foreign Operations Department, Cental Bank of Nigeria, Abuja, 28 July 2009.

Mbakwe, Sophia. Statoil, Abuja, 30 July 2009.

Momoh, Adamson, National Union of Petroleum and Natural Gas Workers (NUPENG), Lagos, 4 August 2009.

Muhammed. Salisu D., Funds Office, Foreign Operations Department, Central Bank of Nigeria, Abuja, 28 July 2009.

Nwadishi, Faith, Publish What You Pay. Abuja, 29 July 2009.

Odey, Sylvia, Economic and Financial Crimes Commission, Abuja, 28 July 2009.

Ofikhenua, John, *The Nation,* Abuja, 6 August 2009.

Olaide, Dayo, Coordinator, West African Resource Watch (WARW), Open Society Institute for West Africa (OSIWA), Abuja, 1 August 2009.

Sa'eed, Haruna Yinusa, Executive Secretary, NEITI, Abuja, 30 July 2009.

Williams, Alabi, Assistant Editor, *The Guardian,* Lagos, 4 August 2009.

B. Organisations represented at focus group discussion at the Centre for Democracy and Development Offices, Abuja on 27 July 2009
Centre for Democracy and Development (CDD)
Environmental Rights Action
Initiative for Community Development
PACT-Nigeria
Poverty Related Diseases College (PRD College)
Publish What You Pay (PWYP)
Real Empowerment
Revenue World Institute (RWI)
Stakeholders Democracy Network (SDN)
Transition Monitoring Group (TMG)
West African Resorce Watch (WARW)
Women Rights Education Centre (WREC)
World Bank
Zero Corruption Coalition (ZCC)

In their *Natural Resource Governance and EITI Implementation in Nigeria*, Musa Abutudu and Dauda Garuba provide the most up-to-date and in-depth analysis of the Nigerian Extractive Industries Transparency Initiative (NEITI), providing a balanced yet critical evaluation of its performance, limitations and potential as an institution for helping Africa's largest oil exporter to escape the so-called resource curse and lay a firm basis for sustainable development. This Current African Issue contains valuable insights and information that will be of interest to all those with a keen interest in institutionalising transparency and accountability in natural resource governance in Africa.

CURRENT AFRICAN ISSUES PUBLISHED BY THE INSTITUTE

Recent issues in the series are available electronically
for download free of charge www.nai.uu.se

1. *South Africa, the West and the Frontline States. Report from a Seminar.* 1981, 34 pp, (out-of print)

2. Maja Naur, *Social and Organisational Change in Libya.* 1982, 33 pp, (out-of print)

3. *Peasants and Agricultural Production in Africa. A Nordic Research Seminar. Follow-up Reports and Discussions.* 1981, 34 pp, (out-of print)

4. Ray Bush & S. Kibble, *Destabilisation in Southern Africa, an Overview.* 1985, 48 pp, (out-of print)

5. Bertil Egerö, *Mozambique and the Southern African Struggle for Liberation.* 1985, 29 pp, (out-of print)

6. Carol B.Thompson, *Regional Economic Polic under Crisis Condition. Southern African Development.* 1986, 34 pp, (out-of print)

7. Inge Tvedten, *The War in Angola, Internal Conditions for Peace and Recovery.* 1989, 14 pp, (out-of print)

8. Patrick Wilmot, *Nigeria's Southern Africa Policy 1960–1988.* 1989, 15 pp, (out-of print)

9. Jonathan Baker, *Perestroika for Ethiopia: In Search of the End of the Rainbow?* 1990, 21 pp, (out-of print)

10. Horace Campbell, *The Siege of Cuito Cuanavale.* 1990, 35 pp, (out-of print)

11. Maria Bongartz, *The Civil War in Somalia. Its genesis and dynamics.* 1991, 26 pp, (out-of print)

12. Shadrack B.O. Gutto, *Human and People's Rights in Africa. Myths, Realities and Prospects.* 1991, 26 pp, (out-of print)

13. Said Chikhi, Algeria. *From Mass Rebellion to Workers' Protest.* 1991, 23 pp, (out-of print)

14. Bertil Odén, *Namibia's Economic Links to South Africa.* 1991, 43 pp, (out-of print)

15. Cervenka Zdenek, *African National Congress Meets Eastern Europe. A Dialogue on Common Experiences.* 1992, 49 pp, ISBN 91-7106-337-4, (out-of print)

16. Diallo Garba, *Mauritania–The Other Apartheid?* 1993, 75 pp, ISBN 91-7106-339-0, (out-of print)

17. Zdenek Cervenka and Colin Legum, *Can National Dialogue Break the Power of Terror in Burundi?* 1994, 30 pp, ISBN 91-7106-353-6, (out-of print)

18. Erik Nordberg and Uno Winblad, *Urban Environmental Health and Hygiene in Sub-Saharan Africa.* 1994, 26 pp, ISBN 91-7106-364-1, (out-of print)

19. Chris Dunton and Mai Palmberg, *Human Rights and Homosexuality in Southern Africa.* 1996, 48 pp, ISBN 91-7106-402-8, (out-of print)

20. Georges Nzongola-Ntalaja *From Zaire to the Democratic Republic of the Congo.* 1998, 18 pp, ISBN 91-7106-424-9, (out-of print)

21. Filip Reyntjens, *Talking or Fighting? Political Evolution in Rwanda and Burundi, 1998–1999.* 1999, 27 pp, ISBN 91-7106-454-0, SEK 80.-

22. Herbert Weiss, *War and Peace in the Democratic Republic of the Congo.* 1999, 28 pp, ISBN 91-7106-458-3, SEK 80,-

23. Filip Reyntjens, *Small States in an Unstable Region – Rwanda and Burundi, 1999–2000,* 2000, 24 pp, ISBN 91-7106-463-X, (out-of print)

24. Filip Reyntjens, *Again at the Crossroads: Rwanda and Burundi, 2000–2001.* 2001, 25 pp, ISBN 91-7106-483-4, (out-of print)

25. Henning Melber, *The New African Initiative and the African Union. A Preliminary Assessment and Documentation.* 2001, 36 pp, ISBN 91-7106-486-9, (out-of print)

26. Dahilon Yassin Mohamoda, *Nile Basin Cooperation. A Review of the Literature.* 2003, 39 pp, ISBN 91-7106-512-1, SEK 90,-

27. Henning Melber (ed.), *Media, Public Discourse and Political Contestation in Zimbabwe.* 2004, 39 pp, ISBN 91-7106-534-2, SEK 90,-

28. Georges Nzongola-Ntalaja, *From Zaire to the Democratic Republic of the Congo.* Second and Revised Edition. 2004, 23 pp, ISBN-91-7106-538-5, (out-of print)

29. Henning Melber (ed.), *Trade, Development, Cooperation – What Future for Africa?* 2005, 44 pp, ISBN 91-7106-544-X, SEK 90,-

30. Kaniye S.A. Ebeku, *The Succession of Faure Gnassingbe to the Togolese Presidency – An International Law Perspective.* 2005, 32 pp, ISBN 91-7106-554-7, SEK 90,-

31. Jeffrey V. Lazarus, Catrine Christiansen, Lise Rosendal Østergaard, Lisa Ann Richey, *Models for Life – Advancing antiretroviral therapy in sub-Saharan Africa.* 2005, 33 pp, ISBN 91-7106-556-3, SEK 90,-

32. Charles Manga Fombad and Zein Kebonang, *AU, NEPAD and the APRM – Democratisation Efforts Explored.* Edited by Henning Melber. 2006, 56 pp, ISBN 91-7106-569-5, SEK 90,-

33. Pedro Pinto Leite, Claes Olsson, Magnus Schöldtz, Toby Shelley, Pål Wrange, Hans Corell and Karin Scheele, *The Western Sahara Conflict – The Role of Natural Resources in Decolonization.* Edited by Claes Olsson. 2006, 32 pp, ISBN 91-7106-571-7, SEK 90,-

34. Jassey, Katja and Stella Nyanzi, *How to Be a "Proper" Woman in the Times of HIV and AIDS.* 2007, 35 pp, ISBN 91-7106-574-1, SEK 90,-

35. Lee, Margaret, Henning Melber, Sanusha Naidu and Ian Taylor, *China in Africa.* Compiled by Henning Melber. 2007, 47 pp, ISBN 978-91-7106-589-6, SEK 90,-

36. Nathaniel King, *Conflict as Integration. Youth Aspiration to Personhood in the Teleology of Sierra Leone's 'Senseless War'.* 2007, 32 pp, ISBN 978-91-7106-604-6, SEK 90,-

37. Aderanti Adepoju, *Migration in sub-Saharan Africa.* 2008. 70 pp, ISBN 978-91-7106-620-6, SEK 110,-

38. Bo Malmberg, *Demography and the development potential of sub-Saharan Africa.* 2008, 39 pp, 978-91-7106-621-3

39. Johan Holmberg, *Natural resources in sub-Saharan Africa: Assets and vulnerabilities.* 2008, 52 pp, 978-91-7106-624-4

40. Arne Bigsten and Dick Durevall, *The African economy and its role in the world economy.* 2008, 66 pp, 978-91-7106-625-1

41. Fantu Cheru, *Africa's development in the 21st century: Reshaping the research agenda.* 2008, 47 pp, 978-91-7106-628-2

42. Dan Kuwali, Persuasive Prevention. *Towards a Principle for Implementing Article 4(h) and R2P by the African Union.* 2009. 70 pp. ISBN 978-91-7106-650-3

43. Daniel Volman, *China, India, Russia and the United States. The Scramble for African Oil and the Militarization of the Continent.* 2009. 24 pp. ISBN 978-91-7106-658-9

44. Mats Hårsmar, *Understanding Poverty in Africa? A Navigation through Disputed Concepts, Data and Terrains.* 2010. 54 pp. ISBN 978-91-7106-668-8

45. Sam Maghimbi, Razack B. Lokina and Mathew A. Senga, *The Agrarian Question in Tanzania? A State of the Art Paper.* 2011. 67 pp. ISBN 978-91-7106-684-8

46. William Minter, *African Migration, Global Inequalities, and Human Rights. Connecting the Dots.* 2011. 95 pp. ISBN 978-91-7106-692-3

47. Musa Abutudu and Dauda Garuba, *Natural Resource Governance and EITI Implementation in Nigeria.* 2011. 75 pp. ISBN 978-91-7106-708-1

www.ingramcontent.com/pod-product-compliance
Lightning Source LLC
Chambersburg PA
CBHW080020280326
41934CB00015B/3421